National Electrical Code®
Study Guide for Electrical
Engineers and Technologists

Solved electrical questions to familiarize electrical engineers and technologists with the 2014 *NEC*®

Stephen P. Tubbs, P.E.
formerly of the
Pennsylvania State University,
currently an
industrial consultant

NOTICE TO THE READER

The author does not warrant or guarantee any of the products, equipment, or programs described herein or accept liability for any damages resulting from their use.

The reader is warned that electricity and the construction of electrical equipment are dangerous. It is the responsibility of the reader to use common sense and safe electrical and mechanical practices.

National Electric Code, NEC, NFPA 70 and NFPA are registered trademarks of the National Fire Prevention Association, Inc., Quincy, MA 02169.

The cover photograph shows work being done on an actual residential panelboard.

Printed in the United States of America

ISBN   978-0-9819753-5-1

CONTENTS

PAGE

INTRODUCTION-- iv

1.0 ORGANIZATION OF THE *NEC* 2014-- 1

    1.1 WHAT IS THE *NEC* 2014?-- 1

    1.2 WHAT IS THE NFPA?-- 1

    1.3 LOCATION OF MATERIAL IN THE *NEC* 2014-------------------------------- 2

    1.4 FINDING THINGS IN THE *NEC* 2014-- 2

    1.5 WHERE TO GET THE *NEC* 2014--- 2

    1.6 HOW DIFFERENT ARE EARLIER *NEC* EDITIONS?--------------------------- 3

    1.7 SIMPLIFIED *NEC* DEFINITIONS--- 3

2.0 EXAMPLE QUESTIONS-- 7

3.0 EXAMPLE QUESTION SOLUTIONS--- 85

4.0 QUESTIONS THE *NEC* DOES NOT ANSWER---------------------------------- 145

5.0 OTHER SOURCES OF INFORMATION-- 149

    5.1 REFERENCES--- 149

    5.2 USEFUL WEBSITE-- 149

    5.3 NFPA--- 149

    5.4 WHERE TO PURCHASE THE *NEC*-- 149

    5.5 *NEC* CLASSES-- 150

9.0 *NEC* ABBREVIATIONS USED IN THIS BOOK------------------------------- 151

INTRODUCTION

The purpose of this book is to familiarize electrical engineers and technologists with the *NEC* (*National Electrical Code*) 2014 edition. It should not be used as a design guide; the reader should carefully check his own circuit designs against the *NEC* rather than against this book.

The reader should obtain a print or electronic copy of the *NEC* to use as he goes through this book's example questions.

The *NEC* is published by the National Fire Prevention Association (NFPA). The *NEC* is also titled *NFPA 70*. It covers the installation of raceways, equipment, electrical conductors, signaling and communications conductors, and optical fiber cables in commercial, residential, and industrial buildings. It has been adopted in all 50 U.S. states.

Many find the *NEC* confusing. Sometimes determining if a circuit is in compliance with the *NEC* can seem more like a courtroom debate over legal interpretations than a technical exercise. At the same time, most agree that the *NEC* is the most important U.S. electrical standard for electrical construction. It is even used as a guide on electrical systems that it was not designed for, such as those on ships and railroad rolling stock.

The *NEC* is primarily studied and used by electricians. They must pass exams on it. Electrical engineers and electrical engineering technologists often receive a more theoretical electrical education and begin their careers with no knowledge of the *NEC*. This book quickly introduces the *NEC* to academically trained electrical engineers and technologists.

This book is written for the *NEC* 2014 edition. The instruction method is the posing and answering of example electrical design questions that would benefit by reference to the NEC 2014. Through searching for solutions in the *NEC*, the learner becomes familiar with it.

Different series of questions are posed for each of the major chapters of the *NEC* 2014. The learner should attempt each question using his knowledge and his copy of the *NEC* 2014. Afterward, the learner's answers should be compared to those given in this book.

In this book, *NEC* 2014 will usually be written simply as *NEC*.

This book's example questions and selection of *NEC* solutions are not authorized or disallowed by the NFPA. They are the work of this author.

1.0 ORGANIZATION OF THE *NEC* 2014

1.1 WHAT IS THE *NEC* 2014?

The *NEC* 2014 is a standard that was published and copyrighted by the NFPA in 2013. It covers the installation of electrical wiring and equipment, usually those used in buildings.

The *NEC* itself is not a legal document. The writers of it have no authority to enforce it. However, many Authorities Having Jurisdiction (AHJ), such as local governments, have adopted it, or part of it, as a required standard. Those authorities then consider it as the law and enforce compliance to it.

The first *NEC* was published in 1897. It has been revised periodically. Currently, a new edition is created every three years.

The best description of what the *NEC* covers is in the *NEC* itself on Page 23[*], Section 90.2, Scope.

1.2 WHAT IS THE NFPA?

NFPA stands for the National Fire Prevention Association. In an NFPA website it is stated, "The mission of the international nonprofit NFPA, established in 1896, is to reduce the worldwide burden of fire and other hazards on the quality of life by providing and advocating consensus codes and standards, research, training, and education."

The NFPA code books cover a wide range of fire related topics. On its website it sells 366 books of codes and standards.

[*] Page numbers in the *NEC* are prefixed by 70-. The 70- prefix will not be used in this book.

In electrical topics, besides the *NEC*, the NFPA publishes and sells:

> *NFPA 70B: Recommended Practices for Electrical Equipment Maintenance*
> *NFPA 70E: Standard for Electrical Safety in the Workplace,*
> *NFPA 73: Standard for Electrical Inspections for Existing Dwellings*
> *NFPA 75: Standard for the Fire Protection of Information Technology Equipment*
> *NFPA 76: Standard for the Fire Protection of Telecommunications Facilities*
> *NFPA 77: Recommended Practice on Static Electricity*
> *NFPA 79: Electrical Standard for Industrial Machinery,*
> *NFPA 110: Standard for Emergency and Standby Power Systems*
> *NFPA 111: Standard on Stored Electrical Energy Emergency and Standby Power Systems*
> *NFPA 731: Standard for the Installation of Electronic Premises Security Systems*
> *NFPA 780: Standard for the Installation of Lightning Protection Systems*
> *NFPA 791: Recommended Practice and Procedures for Unlabeled Electrical Equipment Evaluation*
> *NFPA 850: Recommended Practice for Fire Protection for Electric Generating Plants and High Voltage Direct Current Converter Stations*
> *Electrical Inspection Manual*

1.3 LOCATION OF MATERIAL IN THE *NEC* 2014

The most general material is in *NEC* Chapters 1 to 4 (pages 23 to 381) and in the tables in *NEC* Chapter 9 (pages 756 to 771). These chapters are the ones usually used.

The material in *NEC* Chapters 5 to 8 (pages 382 to 755) deals with special electrical occupancies, equipment, and conditions and communications systems.

The Informative Annexes (pages 772 to 867) are not part of the *NEC* code, but can be very useful. Annex D, for example, gives sample circuit calculations.

1.4 FINDING THINGS IN THE *NEC* 2014

Until recently *NEC* users found pertinent Sections by using their familiarity with the *NEC*, its index, and its table of contents. Now that computer PDF and e-reader versions of the *NEC* are available, relevant Sections can often be found quicker with keyword searches.

1.5 WHERE TO GET THE *NEC* 2014

Copies can be purchased from online bookstores and directly from the NFPA. On the internet the NFPA can be found at http://www.nfpa.org/

1.6 HOW DIFFERENT ARE EARLIER *NEC* EDITIONS?

The *NEC* is continuously changing. The public is encouraged to suggest improvements to it. Suggested improvements are evaluated by committees and accepted or rejected by NFPA member votes. Now the *NEC* 2014 is being reviewed. The new *NEC* 2017 will be different.

The changes from the last *NEC* edition are marked.

1) Large blocks of changed or new material have a vertical line to the left of them. For example, see Page 83, the new figure in Section 230.1.

2) Changed material is highlighted like this with a grey background. For example, see Page 23, the last sentence of Section 90.1 (A).

3) Material that has been moved into the *NEC* from another NFPA document is marked with the document number in square brackets. For example, see Page 453, Section 516.2. Definitions, "Limited Finishing Workstation". At the end of this it is written [**33**:3.3.15.1]. This shows the material came from NFPA 33 Section 3.3.15.1.

4) Removed material is marked with a bullet point, "●". For example, see Page 23, the bullet point between Section 90.1(B) and Section 90.1(C). The Section that was 90.1(C) in the older edition was moved or removed and the old 90.1(D) became the 90.1(C) of the *NEC* 2014.

Even though there are many *NEC* changes, there is much more that has remained the same. A reader who is familiar with a previous *NEC* edition will have little difficulty adapting to the *NEC* 2014.

1.7 SIMPLIFIED *NEC* DEFINITIONS

The *NEC* use of the words such as "bond", "grounded", and "grounding" can be confusing. One *NEC* expert, in a Mike Holt Enterprises Inc. website wrote, "…It took me over 20 years to figure out that when the *NEC*, as well as books and magazine articles, state "ground or grounded" they rarely intend that the metal parts be connected to the earth…" The same website also states, "…the *NEC* often uses the term "ground" when it really means "bond"…".

Probably the confusion came about because the original *NEC* was written over 100 years ago and then was revised by committees.

Below are simplified NEC definitions. More exact *NEC* definitions can be obtained from the *NEC*.

1) <u>Conductor</u> is a substance that allows electricity to pass continuously through it. Usually, we think of conductors as cables and bus bars, but in the *NEC* and the following definitions a <u>Conductor</u> may be a metal item such as metal conduit or cable tray.

2) <u>Grounded (Grounding)</u> is connected (connecting) to earth ground or by a <u>Conductor</u> to earth ground.

3) <u>Ungrounded</u> is not connected to earth ground or by a <u>Conductor</u> to earth ground.

4) <u>Ungrounded Conductor</u> is a <u>Conductor</u> that is usually electrically charged relative to earth ground. Commonly it is called a "hot", "live", "line", or "phase" <u>Conductor</u>. Usually it is a cable or bus bar. See this book's Chapter 3.0 Figure 3-1 (page 83).

5) <u>Neutral Conductor</u> is the conductor connected to the neutral point of a system. The <u>Neutral Conductor</u> is intended to carry current in single-phase circuits and may carry it in 3-phase circuits. Usually the <u>Neutral Conductor</u> is <u>Grounded</u>. When it is <u>Grounded</u>, it may be called both a <u>Neutral Conductor</u> and a <u>Grounded Conductor</u>. See this book's Chapter 3.0 Figure 3-2 (page 85).

6) <u>Grounding Electrode Conductor</u> is a <u>Conductor</u> used to connect the system <u>Grounded Conductor</u> or the equipment to a grounding electrode or to a point on the grounding electrode system. See this book's Chapter 3.0 Figures 3-1 and 3-2 (pages 83 & 85).

7) <u>Separately Derived System</u> is a wiring system whose power is derived from a source of electric energy or equipment other than the electric utility service. Such systems have no direct electrical connection to the electric utility other than grounding and bonding. See this book's Chapter 3.0 Figures 3-2 to 3-5 (page 85 and 90 to 92).

8) <u>Equipment Grounding Conductor</u> is a normally non-current-carrying conductor used to connect metal parts of equipment, raceways, or other enclosures to the <u>System Grounded Conductor</u> and/or <u>Grounding Electrode Conductor</u> at the service entrance or at the <u>Separately Derived System</u> Source. Sometimes it is called a <u>Grounding Conductor</u>. See this book's Chapter 3.0 Figure 3-1 (page 83).

9) <u>Grounded Conductor</u> is a system or conductor that is intentionally <u>Grounded</u>. A <u>Grounded Conductor</u> is not the same as a <u>Grounding Conductor</u>. A <u>Grounded Conductor</u> may carry current under normal operation and may be a <u>Neutral Conductor</u>. See this book's Chapter 3.0 Figure 3-1 (page 83).

10) Grounding Conductor is the same as Equipment Grounding Conductor. It is not the same as a Grounded Conductor. See this book's Chapter 3.0 Figure 3-2 (page 85).

11) Bonding is the electrically connecting of exposed metallic items that are not designed to carry electricity. Usually bonded items are connected by a conductive path (often an Equipment Grounding Conductor) to earth ground. Bonding helps protect against electric shock.

12) Bonding Conductor or Jumper is a Conductor electrically connecting metal parts. See this book's Chapter 3.0 Figure 3-1 (page 83).

13) Supply-Side Bonding Jumper is a Conductor installed on the supply-side of a service for a Separately Derived System. It connects exposed metal parts to ground. See this book's Chapter 3.0 Figures 3-1 and 3-2 (pages 83 and 85).

14) System Bonding Jumper is a Conductor installed between the Grounded Conductor and the Supply-Side Bonding Jumper and/or the Equipment Grounding Conductor at a Separately Derived System. See this book's Chapter 3.0 Figure 3-2 (page 85).

2.0 EXAMPLE QUESTIONS

"It is not the answer that enlightens, but the question." – *Découvertes* – a play by Eugene Ionesco

The solutions for the following example questions are in the *NEC* 2014. For each question the learner should write the appropriate page number(s), Section or Article number(s), and an abbreviated plain English answer into the space provided.

In most cases, the questions require the citing of a single *NEC* Section. Those questions are in the same order as the *NEC* Sections. For example, the answer to question 1 is found on *NEC* Page 23 in Section 90.2, the answer to question 2 is on Page 24 in Section 90.4, and so on. However, some of the questions, such as 3 and 4, require answers from more than one *NEC* Section. Their answers cannot be in the same order as the *NEC* Sections.

Should one go through the questions in numerical order, 1, 2, 3, etc? The reader should decide. Some would appreciate that the location of the *NEC* Section answer to one question is near to that of the previous question. Others would rather skip around through this book's questions so that the relevant *NEC* answers are scattered, as they are in real life. The reader will learn about the *NEC* either way.

Answers to the example questions are in this book's Chapter 3.0. Use Chapter 3.0 answers only after thoroughly searching the *NEC*.

1) What installations are not covered by the *NEC*?

2) Is the *NEC* intended to be suitable for mandatory application by governments?

3) Put the appropriate descriptor for each letter in Figure 2-1 into the following table. (Information is required from several Sections.)

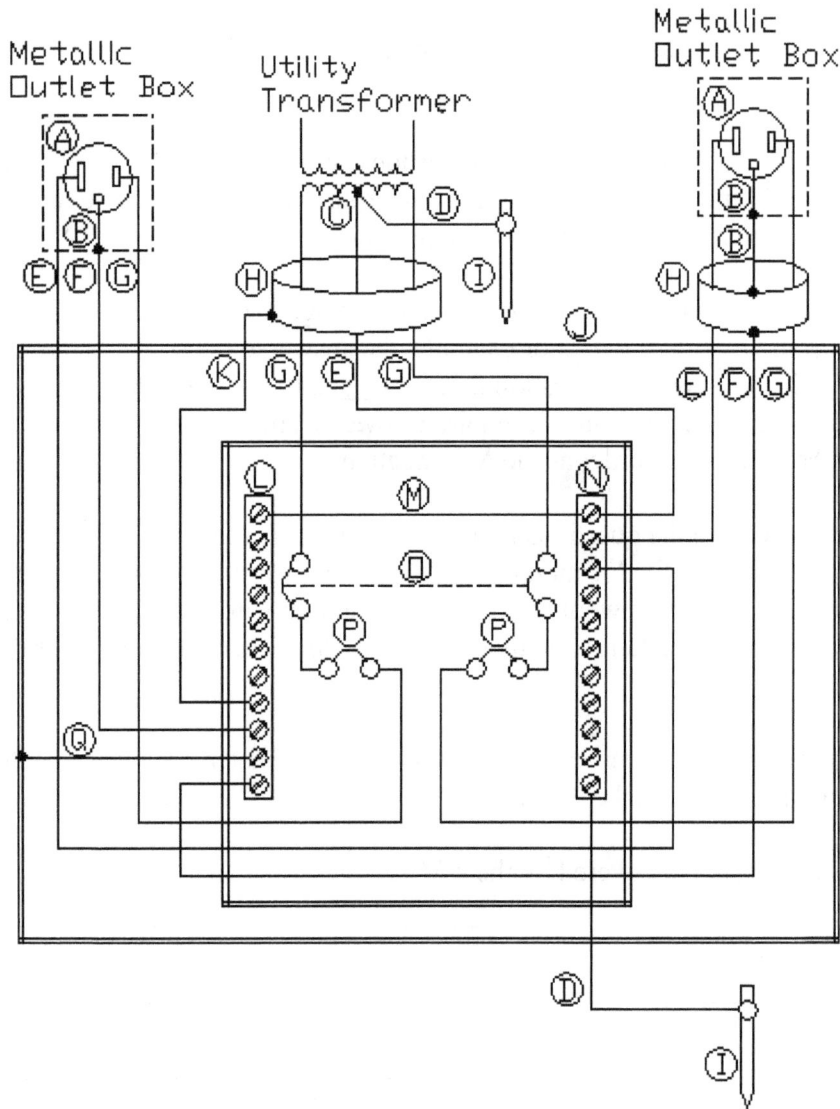

Panelboard (Service Panel)	Receptacle	Bonding Conductor
Equipment Bonding Jumper	Circuit Breaker	Grounding Electrode
Equipment Grounding Conductor	Grounded Conductor (Neutral Conductor)	Service Disconnect (Main Circuit Breaker)*
Grounding Electrode Conductor	Neutral Bus*	Main Bonding Jumper
Neutral Point	Ungrounded Conductor*	Ground Bus*
Metallic Raceway (Conduit)*	Supply-Side Bonding Jumper	

*Note: This term is not precisely defined in the *NEC*. The term is included here because it is useful and it appears in the *NEC* text.

Figure 2-1 Service entrance and panel *NEC* descriptors.

Letter	Descriptor	*NEC* Page	Section Number
A			
B			
C			
D			
E			
F			
G			
H			
I			
J			
K			
L			
M			
N			
O			
P			
Q			

4) Write the appropriate descriptors into the Figure 2-2. (Information is required from several Sections.)

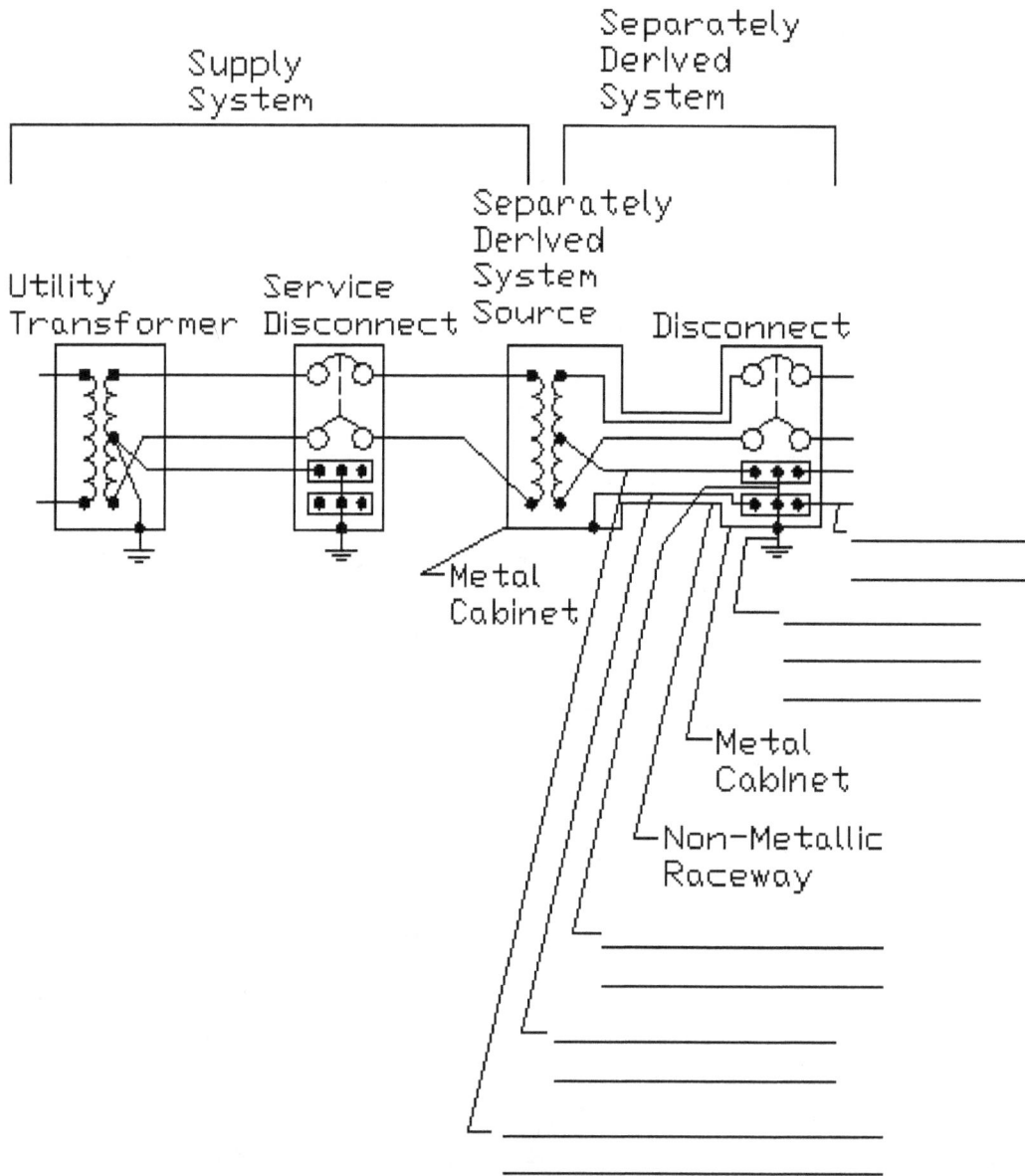

System Bonding Jumper	Grounding Electrode Conductor
Grounding Conductor	Grounded Conductor (Neutral)
Supply-Side Bonding Jumper	

Figure 2-2 Bonding and Grounding details of a Separately Derived System.

5) Define the following:

a) Hard Conversion and Soft Conversion

b) Ampacity

c) Authority Having Jurisdiction (AHJ)

d) Bonded

e) Bonding Conductor or Jumper

f) Bonding Jumper, System

g) Branch-Circuit

h) Circuit Breaker

i) Explosion Proof

j) Feeder

k) Grounded Conductor

l) Ground-Fault

m) Grounding Conductor.

n) Interrupting-Rating

o) Listed

p) Outlet

q) Overcurrent

r) Overload

s) Panelboard

t) Qualified Person

-16-

u) Raceway

v) Separately Derived System

w) Service

x) Service Equipment

y) Short-Circuit Current Rating

z) Ungrounded

6) What is the minimum time for a *NEC* "continuous load"?

7) What is an intersystem bonding termination device used for?

8) What is the minimum Interrupting-Rating for Service, Feeder, and Branch Overcurrent Devices?

9) Does the *NEC* require that listed or labeled equipment be installed and used in accordance with instructions included in the listing or labeling?

10) What words does the *NEC* use to state that electrical equipment and wiring be neatly and conventionally installed?

11) May unused openings to electrical equipment be left open?

12) May conductors of dissimilar metals be intermixed in a terminal where the metals have contact with each other?

13) Does the *NEC* allow house wiring to be extended by simply stripping, joining by twisting, and insulating with tape additional lengths of wire? (Information is available from several Sections.)

14) It is often difficult for an electrical inspector to prove that an electrical installation was not made in a "neat and workmanlike manner". However, it is easier to find other code violations. Cite a few common code violations. (Information is available in several Sections.)

15) Does the *NEC* allow conductors to have higher temperature ratings than their terminations?

16) Where should Arc-Flash warnings be posted?

17) A 480 volts ac safety switch box is to be installed in a corridor, as shown in the following drawing. The walls are concrete. What are the minimum dimensions allowed by the *NEC*? (Information is required from a Section and Table.)

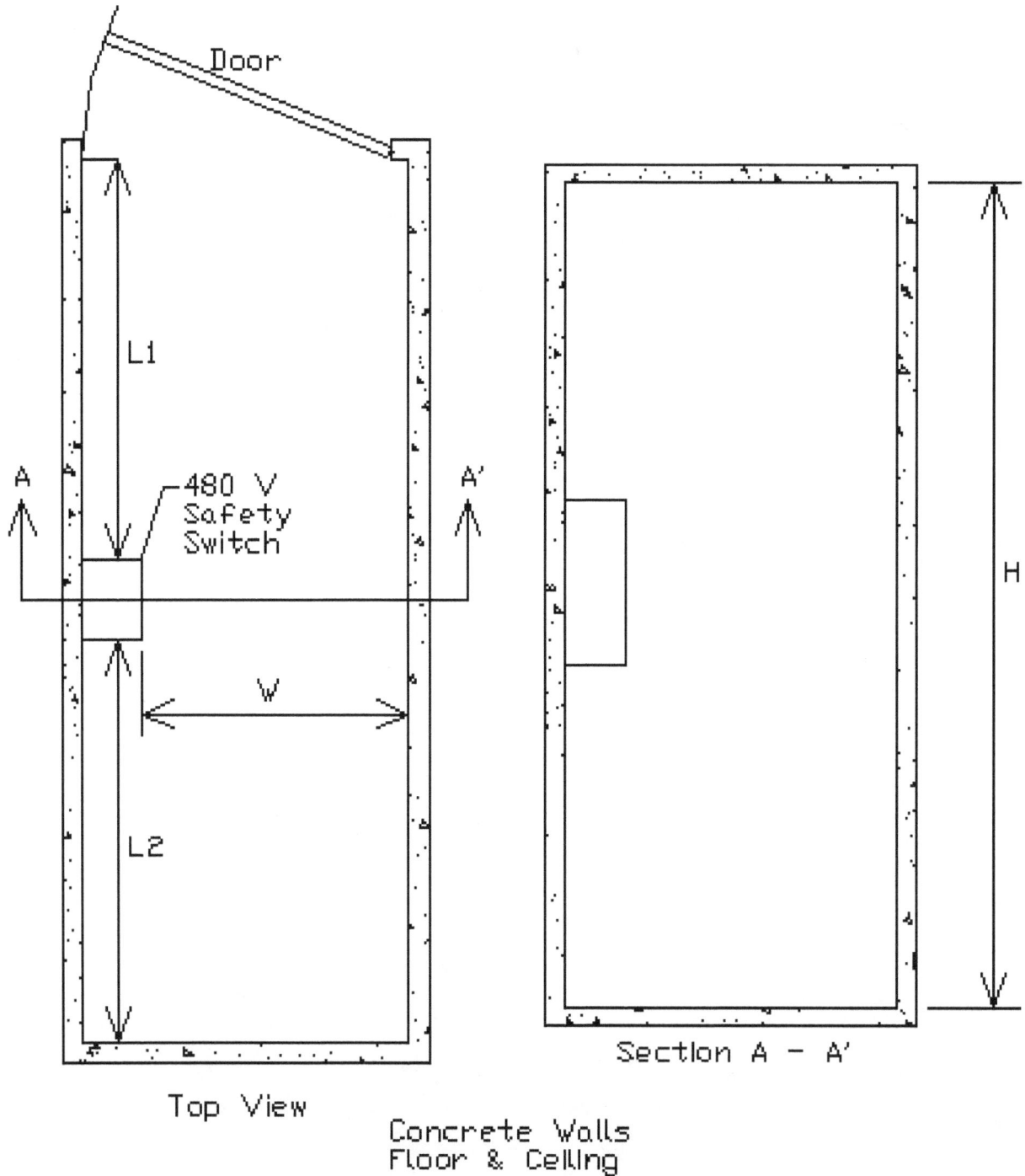

Figure 2-3 480 volts safety switch working space.

18) Would the *NEC* allow the storage of boxes under the safety switch of Figure 2-3?

19) When is an electrical room required to have two exits?

20) According to the *NEC*, when must live parts be guarded against accidental contact by the use of approved enclosures or by other *NEC* approved means?

21) What are the different indoor and outdoor enclosure types?

22) Are equipment grounding conductors required to be bare or insulated?

23) May a grounded conductor (neutral conductor) depend on a metal enclosure or metal cable armor for continuity?

24) May neutral conductors be used for more than one branch-circuit?

25) What conductors use white insulation?

26) What are the current ratings of multi-outlet branch-circuits?

27) What Specific–Purpose Branch-Circuits are described in the *NEC*?

28) How do you find the Overcurrent Device ratings and conductor ratings needed for X-ray equipment? (Information can be found in a Table and Section)

29) What colors does the *NEC* specify for dc ungrounded conductors smaller than 6 AWG?

30) Where are GFCIs (Ground-Fault Circuit Interrupters) required?

31) Where are AFCIs (Arc Fault Circuit Interrupters) required?

32) What is the minimum ampacity of a branch-circuit conductor? (Information is required from several Sections.)

33) Does the *NEC* specify the percent voltage drop from the service to the load for circuits that have voltages not more than 600 volts? (Information is available from several Sections.)

34) A branch-circuit supplies a continuous load of 10 amperes and an intermittent load of 15 amperes. What Overcurrent Device protection rating should be selected? If THHN copper conductor is to be used at an ambient temperature of 30°C what gauge should be selected? (Information is required from a Section and several Tables.)

| |
| |
| |
| |
| |
| |

35) What receptacle ampere ratings are required in single receptacles?

| |
| |
| |

36) What are standard current ratings for receptacles in multi-receptacle branch-circuits?

| |
| |
| |

37) How many small appliance receptacles are required for a kitchen countertop?

38) What Section provides general and specific information on feeder circuit cable ampacities?

39) Is feeder Ground-Fault Protection required?
a)   If line to line voltage is 480 volts, it is an ungrounded Δ system, and the disconnect is rated for 1500 amperes.
b)   If line to line voltage is 480 volts, it is a solidly grounded Y system, and the disconnect is rated for 1500 amperes.
c)   If line to line voltage is 208 volts, it is a solidly grounded Y system, and the disconnect is rated for 900 amperes.

40) A school has a floor space of about 4,924 m² (53,000 ft²). What would be the expected lighting load in kilovolt-amperes?

41) What Article covers outside branch and feeder circuits?

42) May cable carrying 1200 volts be installed outside and above ground in flexible conduit? (Information is available in several Sections.)

43) Where is it stated that a means of disconnecting a service is required where ungrounded conductors enter a building?

44) How many Service Disconnects are allowed in one location? (Information is available from several Sections.)

45) A building that has more than one service must have a plaque placed at each service denoting all other services. Where is this stated in the *NEC*? (Information is available from several Sections.)

46) What *NEC* figure graphically shows the relative locations of the Serving Utility, Service-Entrance Conductors, Service Equipment, and Branch-Circuit Feeders?

47) What *NEC* requirements apply to a building that receives electrical power from multiple sources?

48) Usually a building may have only one service (supply). What are some exceptions where more than one supply is allowed?

49) If a Service Conductor is to have a continuous load of 100 amperes and a noncontinuous load of 50 amperes what is its minimum ampacity?

50) Where does the *NEC* state that Service-Entrance Conductors need to be protected against physical damage?

51) What is a Service Disconnect?

52) Does the *NEC* allow the home owner to install a fuse box in his shower?

53) Is access to each occupant's own Service Disconnecting Means required in multiple-occupancy buildings?

| |
| |
| |
| |

54) If Service Conductors supply five Service Disconnect/Overload Devices, each set to 200 amperes, but the calculated load is 900 amperes, what is the minimum ampacity of the Service Conductors?

| |
| |
| |
| |
| |

55) What is a Tap Conductor?

| |
| |
| |
| |

56) Where does the *NEC* state that conductors shall be protected against overcurrent?

57) Do flexible cords, i.e. lamp cords, require overcurrent protection?

58) Does the *NEC* allow pennies to be used in place of fuses in a fuse box or circuit breakers to be taped to the ON position in a load panel?

59) Are Overload and Short-Circuit Protection required on crane lifting electromagnets?

60) When sizing overcurrent protection, sometimes the Overcurrent Protective Device current rating should be the next higher from the calculated current and sometimes it should be the next lower. Find examples of this in the *NEC*.

| |
| |
| |
| |
| |
| |

61) What are the maximum Overcurrent Device ratings for 18, 16, 14, 12, and 10 AWG copper wire?

| |
| |
| |

62) When does the *NEC* consider extension cords protected?

| |
| |
| |

63) What are some standard fuse and circuit breaker ratings?

64) If an adjustable setting circuit breaker does not have restricted access to its adjustment how must it be used?

65) May ordinary single fuses be connected in parallel?

66) What conductors shall have Overcurrent Protection?

67) Where should a conductor's Overcurrent Protection be located?

68) What is a major difference between the 0 to 3 m (0 to 10 ft) and >3 to 7.5 m (>10 to 25 ft) feeder tap rules?

69) Is it necessary to place Overcurrent Protection on a transformer's secondary? (Information is required from several Sections.)

70) When does the *NEC* require arc energy reduction?

71) What is the Article on grounding?

72) What are the reasons for grounding?

73) What is an effective Ground-Fault current path?

74) May an outdoor receptacle be grounded locally through its own ground rod?

| |
| |
| |

75) May bonding and grounding connections be made with solder?

| |
| |
| |

76) Where does the *NEC* state that a 120/240 volts ac system, such as that used in a house, shall be grounded?

| |
| |
| |

77) What markings must be put on a 50 to 1000 volts ungrounded ac system?

| |
| |
| |

78) What circuits should not be grounded?

79) May a grounded conductor (neutral conductor) be reconnected to ground at its load equipment?

80) 3/0 AWG copper wire is used for the ungrounded Service Conductors on a 240/120 volts ac system. The Service Conductors are in a single conduit. What is the minimum size for a copper wire grounded Service Conductor? (Information is available from a Section and Table.)

81) In the Separately Derived System of Figure 2-2 the transformer of the Separately Derived System Source does not have its output mid-tap connected directly to its metal cabinet. This is to avoid creating a parallel conducting path ground loop.
a) Where does the *NEC* state that parallel paths should be avoided?
b) If the ungrounded conductor is 3/0 AWG copper wire what is the minimum size for the system bonding jumper and the supply-side bonding jumper?
   (Information is required from a Section and Table.)

| |
| |
| |
| |

82) If parallel ungrounded conductors are run in parallel raceways in a Separately Derived System, are parallel grounded conductors required in each raceway?

| |
| |
| |
| |

83) May a common tapped grounding electrode conductor be used with multiple Separately Derived Systems?

| |
| |
| |
| |

84) What is the main grounding rule for two or more buildings supplied by feeders or branches of the same system?

85) Is a portable standby generator's frame required to be grounded?

86) When is a high impedance neutral to ground allowed?

87) Should electrodes be bonded together to form a grounding electrode system? (Information is required from several Sections.)

| |
| |
| |
| |

88) What length of a metal water pipe must be in contact with the earth for it to be used as a grounding electrode?

| |
| |
| |

89) Are gas pipes and aluminum permitted to be grounding electrodes?

| |
| |
| |

90) What is the maximum resistance of a single rod, pipe, or plate grounding electrode to ground before a supplemental electrode is required?

| |
| |
| |

91) A grounding electrode shall have at least 2.44 m (8 ft) of its length in contact with the soil. What are the alternatives to driving the grounding electrode straight down?

| |
| |
| |

92) Are splices allowed on grounding electrode conductors?

| |
| |
| |

93) If ungrounded service entrance conductors are 1/0 AWG copper, what is the minimum size of the copper grounding electrode conductor?

| |
| |
| |

-46-

94) By what methods may a grounding electrode conductor be connected to its grounding electrode?

95) What must be the current capacity of a bonding jumper?

96) Are standard conduit lock nuts sufficient for bonding?

97) What methods of bonding are acceptable for use in hazardous locations? (Information is required from several Sections.)

98) What may a bonding jumper be made of?

99) A branch-circuit is protected by a 30 amperes breaker. What should be the size of the aluminum bonding jumpers used in this load-side circuit? (Information is required from several Sections.)

100) Ungrounded conductors in an ac service are 3/0 AWG copper. What is the minimum gauge of a copper Grounded Conductor, Main Bonding Jumper, System Bonding Jumper, and Supply-Side Bonding Jumper?

| |
| |
| |
| |

101) Should gas pipes and structural steel be bonded?

| |
| |
| |
| |

102) Is grounding required for a permanently located metal lathe that uses 460 volts ac 3-phase?

| |
| |
| |
| |

103) When should equipment metal parts be connected to an equipment grounding conductor? (Information is required from several Sections.)

104) Name some instances where equipment grounding is not required.

105) What is the maximum length and current capacity of listed flexible metal conduit that is being used as a ground-fault current path?

106) Describe the coverings required for grounding conductors.

107) The color codes in common use for U.S. ac house wiring are:

Conductor Insulation Color Codes	
Color	**Function**
Black	Ungrounded conductor (hot wire)
Red	Ungrounded conductor (hot wire)
Blue	Ungrounded conductor (hot wire)
White coded black[*]	Ungrounded conductor (hot wire)[*]
White	Grounded conductor (neutral wire)
Green	Grounding conductor
Green & yellow stripe	Grounding conductor
Bare copper	Grounding conductor
[*]Usually white wires are grounded. However, occasionally a white wire used in a switch loop will need to be ungrounded. In that case it is marked with a black tape or a spot of black paint at its ends.	

Are these color codes in the *NEC*? (Information is required from several Sections.)

108) Conductors with green or green with yellow stripes insulation are usually equipment grounding conductors. Where could green and green and yellow stripe insulated conductors be used for other than equipment grounding?

| |
| |
| |
| |

109) How close to the earth may aluminum equipment grounding conductors be terminated?

| |
| |
| |

110) Are there cases where the equipment grounding conductor's gauge should be larger than that of the ungrounded conductors?

| |
| |
| |

111) What gauge should the grounding conductor be in a small flexible cord?

112) What is the minimum size of the copper grounding conductor for equipment protected by a 20 amperes circuit breaker?

113) May the structural metal frame of a building be used as ac equipment grounding?

114) When may a receptacle have an isolated ground, a ground that is connected to the grounding electrode conductor but not to its receptacle enclosure?

115) What dc systems need to be grounded?

116) What is the minimum percent ampacity the neutral conductor may have compared to the phase conductors?

117) What is the selection of a metal oxide surge arrestor based on?

118) What are the "Wiring Methods and Materials" referred to in Article 300?

119) The *NEC* says conductors of the same circuit should be kept together. Where is this stated in the *NEC*?

| |
| |
| |

120) May conductors rated to carry more than 1000 volts be next to those rated to carry less than 1000 volts?

| |
| |
| |

121) May active cables and conductors attached to ceiling joists be used as clotheslines? (Information is available in several Sections.)

| |
| |
| |
| |
| |

122) If cables are to go through holes bored in joists, rafters, or wood members, how far back from the surface should the holes be?

123) How deep must conduit containing 120 volts ac wiring be buried beneath a concrete driveway?

124) What sort of conduit would be appropriate for underground use where the conduit may be subject to physical damage?

125) Are splice boxes required on buried cable splices? (Information is available from several Sections.)

126) Does the *NEC* require stainless steel raceways to have a protective coating, as it does for other ferrous metal raceways?

127) May support wires be used as the sole support for hanging conduit from a ceiling?

128) What precaution should be taken when installing electrical conduit and cable to reduce the possibility of fire spread?

129) May wiring be placed in ducts that carry dust, loose stock, or vapors?

130) Where is there general information on conductor designations, insulations, markings, mechanical strengths, ampacity ratings, and uses?

131) May any insulated conductor in the *NEC* be used in dry locations?

132) May conductors be used in ambient temperatures greater than their rated temperature?

133) What are the principal determinants of conductor temperature?

134) If 30 conductors are contained in a conduit how much should the conductors' ampacities be reduced from that for 3 conductors in a conduit?

135) How many °C would the effective ambient temperature be increased on conduit that is mounted outside 380 mm (15 in.) above a roof?

136) Two 12 AWG THHN copper conductors are to be used in conduit at an ambient temperature of 10°C. What is the maximum ambient temperature that these conductors may be used in? What is its current rating at its maximum ambient temperature? What is its current rating at 10°C? (Information is required from several Tables and a Section.)

137) What *NEC* table gives the maximum number of conductors that may be stuffed into a metal box?

138) What is the minimum length for a straight pull box connected to a Metric Designator 27 (Trade Size 1) EMT conduit?

| |
| |
| |
| |

139) What are SE and USE cables?

| |
| |
| |
| |

140) May type UF cable be used as underground service-entrance cable?

| |
| |
| |
| |

141) Where is PVC conduit not to be used?

| |
| |
| |
| |

142) What is a wireway? (Information is required from two Sections.)

143) What maximum percent fill is allowed for metal wireways?

144) Where do you find ampacities for flexible cords?

145) May extension cords be used as a substitute for a structure's fixed wiring?

146) Are splices allowed on flexible cords? (Information is required from a Section and Table.)

147) A fixture is anything permanently attached to a building. A wall-mounted ac electric clock would be an example of an electrical fixture. An electrical fixture is powered by fixture wires. What are the largest and smallest wire gauges the *NEC* allows for fixture wires? (Information is required from a Table and Section.)

148) Where are switches discussed in the *NEC*?

149) How high may the operating handle of a switch or circuit breaker be above the floor?

150) How is an isolated ground receptacle identified?

151) Where does the *NEC* state that receptacles in damp and wet areas must have protection against moisture?

152) What is the maximum Overcurrent Device rating for a feeder to a 400 amperes panelboard? Where must the Overcurrent Device be mounted?

153) Are open or bare incandescent bulb fixtures allowed in closets?

154) Is it necessary to leave open space around recessed lights, or may insulation be put next to them?

155) Is ground fault protection required with a house's gutter de-icing cable?

156) What is the *NEC* Article on motors? What table gives *NEC* Articles and Sections that have information on the application of electric motors in specific equipment? (Information is required from an Article and Table.)

157) Often, *NEC* table values of a motor's full-load current are used rather than the motor's nameplate Full-Load Amperage (FLA) rating values for determining the ampacity of conductors and the ampere ratings of switches and branch-circuit Short-Circuit and Ground-Fault Protection. Where is this stated in the *NEC*? (Information is required from several Sections and Tables.)

158) Where does the *NEC* describe motor nameplate data?

159) What nameplate data should be on a motor controller?

160) Where does the *NEC* require motors to be protected against liquids and dust? (Information is required from several Sections.)

161) What percent of motor full-load nameplate current should a separate Overload Device be rated for the following?
a)   2 horsepower, Service Factor 1.0, temperature rise 40°C
b)   2 horsepower, Service Factor 1.15, temperature rise 30°C
c)   2 horsepower, Service Factor 1.15, temperature rise 40°C
d)   1/2 horsepower, Service Factor 1.15, temperature rise 40°C
    (Information is required from several Sections.)

162) What size dual element fuse is required by a 3-phase induction motor?

163) May three ¼ horsepower 120 volts single-phase induction Thermal Overload Protected motors be powered through the circuit breaker of one branch-circuit?

164) Is a motor controller required to have its own Short-Circuit and Ground-Fault Protective Devices?

165) What protection is allowable on a 120/24 V 40 volt-amperes transformer that is part of a motor controller and in its motor controller enclosure?

| |
| |
| |
| |

166) What size motor controller is needed for a 50 horsepower 460 volts 3-phase induction motor?

| |
| |
| |
| |

167) May motor controllers be used on voltages other than their rated voltage?

| |
| |
| |
| |

168) Is an individual disconnecting means required for each motor controller?

| |
| |
| |
| |

169) May a motor controller disconnect also serve as a motor disconnect?

170) Where does the *NEC* state that motor and motor controller disconnecting means shall plainly indicate open (off) and closed (on) positions?

171) If the rated input current of an inverter is 100 amperes, what is the minimum ampacity of the conductors supplying it?

172) May overload equipment be included within power conversion equipment?

173) Is motor over-temperature protection required if the motor is powered by an adjustable speed drive?

| |
| |
| |
| |

174) A full-wave rectifier supplies a 120 volts dc 1 horsepower motor in a 30°C or less environment.
a) What is the motor's full-load current?
b) What size THHN copper conductors could be used here?
   (Information is required from Sections and Tables.)

| |
| |
| |
| |
| |
| |
| |
| |
| |

175) What gauge SOOW copper 16/3 flexible cord should be used to power a single-phase 1 horsepower 115 V table saw motor? The motor's nameplate full-load amperage is 11 amperes. The table saw will need to be able to operate continuously in a 40 °C environment. The circuit will be protected by a dual element fuse. (Information is required from a Section and several Tables.)

| |
| |
| |
| |
| |
| |
| |
| |
| |
| |
| |
| |
| |
| |
| |
| |

176) What are the full-load and locked rotor currents for a 7½ horsepower, 230 volts single-phase, code letter G induction motor? (Information is required from several Tables.)

177) What current rating should a thermal protector have on a 120 volts 1½ horsepower single-phase motor? (Information is required from a Section and Table.)

178) What gauge copper THHN wire is needed to continuously supply three 230 volts 3-phase induction motors? The motors are 10 horsepower, 3 horsepower, and 1 horsepower. Assume an ambient temperature of 30°C. (Information is required from a Section and several Tables.)

179) What minimum ampere rating is required for an inverse-time circuit breaker used on a 50 horsepower 460 volts 3-phase induction motor? Note: The minimum ampere rating is not the same as overcurrent protection rating. Here ampere rating indicates the capacity of the circuit breaker to carry continuous current. (Information is required from a Section and Table.)

180) The *NEC* states that it covers the installation of all transformers, but then gives exceptions. What are some of the exceptions?

181) What conductor sizes and Overcurrent Protection are needed by an air conditioner that has the following electrical nameplate data? Assume the temperature is no more than 30°C, the conductors are THHN, and there are no more than two current carrying conductors in a raceway. (Information is required from a Section and Table.)

Nameplate of an Air Conditioner that uses an Hermetic Refrigerant Motor-Compressor				
Volts AC	**Min Volts**	**Max Volts**	**Phase**	**Frequency**
208/230	197	253	1	60 Hertz

Minimum Circuit Ampacity	**Max Fuse or Circuit Breaker Amperes**
35.7	50

Compressor		
RLA Amperes	**LRA Amperes**	**Horsepower**
27.1	144	-

Fan		
FLA Amperes	**LRA Amperes**	**Horsepower**
1.9	3.7	1/3

182) What size fuses should a single-phase 100 kilovolt-amperes 2400/240 volts distribution transformer use? Assume an unsupervised location and a percent transformer impedance of less than 6%. (Information is required from a Section and Table.)

183) What size fuses should a single-phase 10 kilovolt-amperes 480/120 volts transformer use? Assume the transformer circuit does not meet the "only primary needs to be protected" specifications of Page 99, Section 240.21 Location in Circuit., (C) Transformer Secondary Conductors. (Information is required from several Tables and a Section.)

184) Do Overcurrent Devices for transformers protect their conductors?

185) What transformers are required to have a disconnecting means? In what locations are the disconnects required?

186) What is a Class I Division 1 location?

187) What does Class III refer to?

188) In a hospital anesthetizing area, what is the highest voltage allowed between conductors before connection to an equipment grounding conductor is required?

189) For the purposes of the bonding of equipotential surfaces in agricultural buildings, are chicken barns and cattle barns required to meet the same bonding requirements?

190) Does the *NEC* allow boats to be supplied power via temporary wiring?

191) What is the longest time that temporary holiday wiring is permitted?

192) Are GFCIs required for temporary construction wiring?

193) The *NEC* has an Article on Information Technology Equipment. What is the NFPA document the *NEC* refers to for the protection of information technology equipment?

194) When must abandoned cables be removed from information technology equipment systems?

195) Is bonding required for swimming pool underwater lighting?

196) A transfer switch should be used when connecting an optional standby generator to a house's electrical system. Where does the *NEC* state this?

197) What are Class 1 circuits? (Information is required from several Sections.)

198) What are Class 2 and Class 3 circuits?

199) Power or lighting conductors may not be in the same raceway as Class 2 or Class 3 conductors. Where is this stated in the *NEC*?

| |
| |
| |

200) Must fire alarm circuits be separated from power and lighting circuits?

| |
| |
| |

201) Determine the minimum size for EMT conduit that contains 8 conductors, each made of stranded 12 AWG THHN cable. (Information is required from several Tables.)

| |
| |
| |
| |
| |
| |

202) What is compact stranding?

3.0 EXAMPLE QUESTION SOLUTIONS

Abbreviated solutions are given here. Abbreviation makes them different from the more complete solutions found in the *NEC*. In many instances the *NEC* offers a general solution and then offers exceptions for specific situations. Always refer to the *NEC* for the best answer. Also note that the same material is often discussed in several Sections in the *NEC*.

In the following "*NEC*" refers to the "*NEC* 2014" and the *NEC* page numbers do not use the 70- prefix that appears in the *NEC* 2014.

1) What installations are not covered by the *NEC*?

<u>Page 23</u>, <u>Section 90.2 Scope.</u>, <u>(B) Not Covered.</u> --- Some installations the *NEC* does not cover are in ships, railroad rolling stock, mines, communications equipment, and electric utility equipment.

2) Is the *NEC* intended to be suitable for mandatory application by governments?

<u>Page 24</u>, <u>Section 90.4 Enforcement.</u> --- Yes.

-86-

3) Put the appropriate descriptor for each letter in Figure 2-1 into the following table. (Information is required from several Sections.)

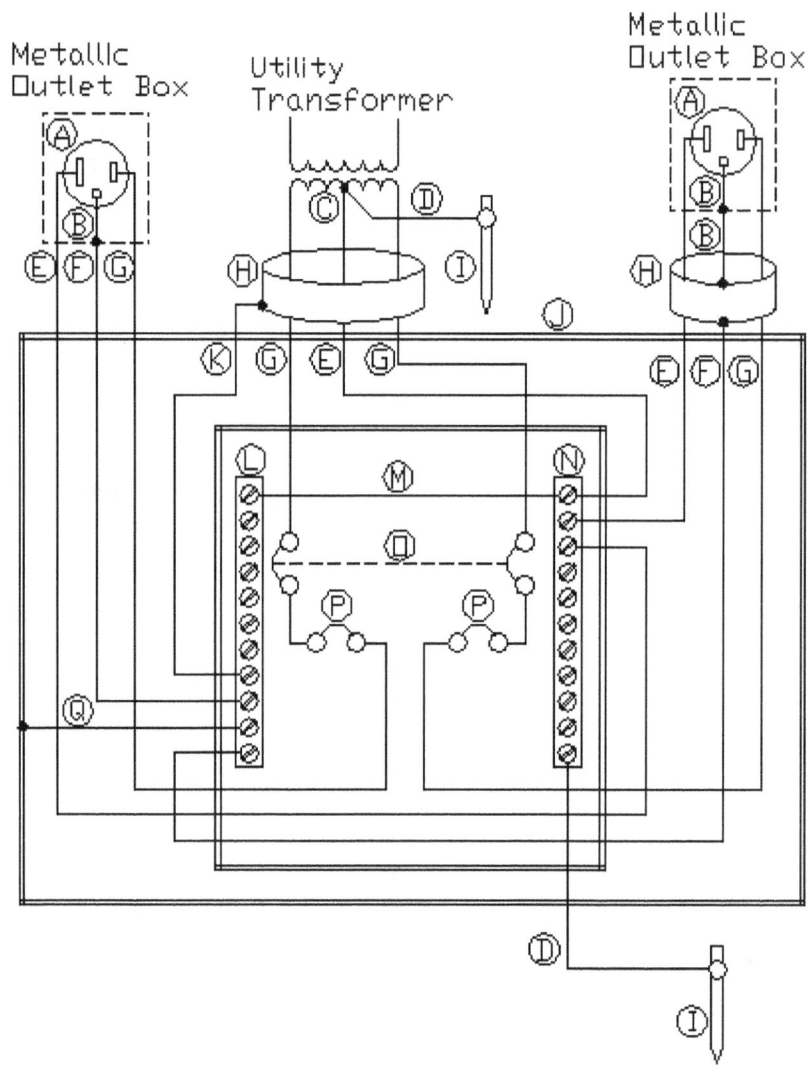

Panelboard (Service Panel)	Receptacle	Bonding Conductor
Equipment Bonding Jumper	Circuit Breaker	Grounding Electrode
Equipment Grounding Conductor	Grounded Conductor (Neutral Conductor)	Service Disconnect (Main Circuit Breaker)*
Grounding Electrode Conductor	Neutral Bus*	Main Bonding Jumper
Neutral Point	Ungrounded Conductor*	Ground Bus*
Metallic Raceway (Conduit)*	Supply-Side Bonding Jumper	

*Note: This term is not precisely defined in the *NEC*. The term is included here because it is useful and it appears in the *NEC* text.

Figure 3-1 Service entrance and panel *NEC* descriptors. (This is the same as Figure 2-1).

Letter	Descriptor	*NEC* Page	Section Number
A	Receptacle	33	100
B	Equipment Bonding Jumper	28	100
C	Neutral Point	32	100
D	Grounding Electrode Conductor	31	100
E	Grounded Conductor (Neutral Conductor)	30	100
F	Equipment Grounding Conductor	30	100
G	Ungrounded Conductor	-----	-----
H	Metallic Raceway (Conduit)	-----	-----
I	Grounding Electrode	31	100
J	Panelboard (Service Panel)	33	100
K	Supply-Side Bonding Jumper	106	250.2
L	Ground Bus	-----	-----
M	Main Bonding Jumper	28	100
N	Neutral Bus	-----	-----
O	Service Disconnect (Main Circuit Breaker)	-----	-----
P	Circuit Breaker	28	100
Q	Bonding Conductor	28	100

4) Write the appropriate descriptors into the Figure 2-2. (Information is required from several Sections.)

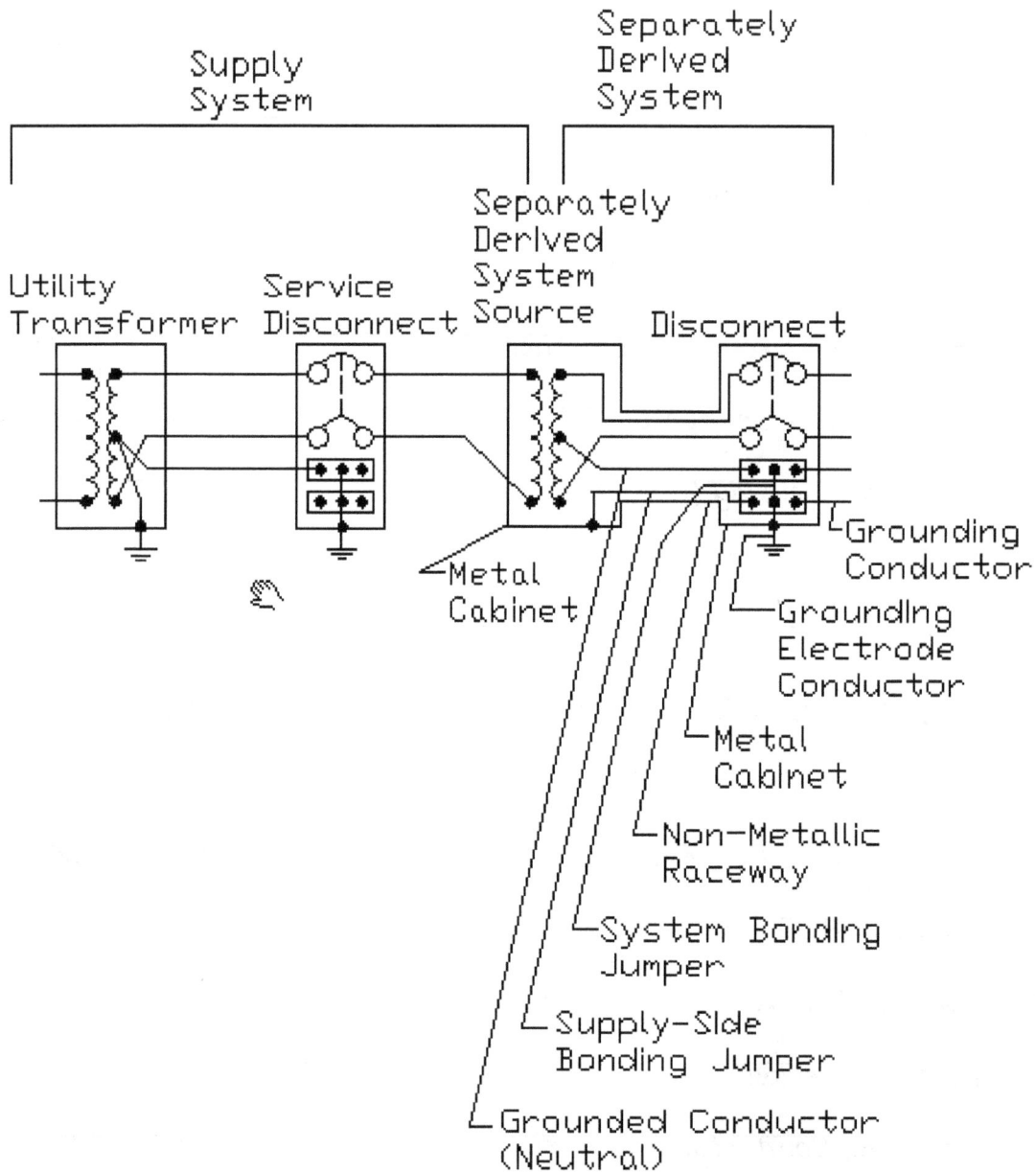

System Bonding Jumper	Grounding Electrode Conductor
Grounding Conductor	Grounded Conductor (Neutral)
Supply-Side Bonding Jumper	

Figure 3-2 Bonding and Grounding details of a Separately Derived System.

5) Define the following:

a) Hard Conversion and Soft Conversion

Page 26, *Section 90.9 Units of Measurement.*, *(D) Compliance.*, *Informational Note No. 1:* --- These refer to the conversion of measurement units from SI to inch-pound and visa-versa. Contrary to what seems logical, a hard conversion is more approximate than a soft conversion. For example, with a hard conversion 6 mm is said to be the same as .25 in. However, with a soft conversion, 6 mm converts to the more exact 6/25.4 = .236 in.

b) Ampacity

Page 27, *Article 100 Definitions.*, *I General.* --- Maximum current that a conductor may carry without exceeding temperature ratings.

c) Authority Having Jurisdiction (AHJ)

Page 27, *Article 100 Definitions.*, *I General.* --- An organization, such as a local government, that chooses to enforce a code. Such an organization could require compliance with the *NEC* with stated exceptions.

d) Bonded

Page 28, *Article 100 Definitions.*, *I General.* --- Electrically connected. Bonded is used to describe the electrical connection of metallic items that are not designed to normally carry electricity but may carry it during a fault. Usually bonded connections have a conduction path to ground. Examples of bonded connections are the connection of a metal conduit to a metal junction box and the conductor (bonding jumper) connection of a receptacle's ground screw to the metal box the receptacle is mounted in.

e) Bonding Conductor or Jumper

Page 28, *Article 100 Definitions.*, *I General.* --- A conductor, bare or insulated, that electrically connects metal parts.

f) Bonding Jumper, System

Page 28, Article 100 Definitions., I General. --- Connection between the Grounded Conductor (Neutral Conductor) and the Supply-Side Bonding Jumper, or the Equipment Grounding Conductor, or both, at a Separately Derived System. An example of a System Bonding Jumper can be seen in Figure 3-2.

g) Branch-Circuit

Page 28, Article 100 Definitions., I General. --- Circuit conductors between the final Overcurrent Device protecting the circuit and the outlets.

h) Circuit Breaker

Page 28, Article 100 Definitions., I General. --- Device designed to open and close a circuit by non-automatic means and automatic overcurrent sensing.

i) Explosion Proof

Page 30, Article 100 Definitions., I General. --- "...capable of withstanding an explosion..."

j) Feeder

Page 30, Article 100 Definitions., I General. --- Circuit conductors between the service equipment and the final branch-circuit Overcurrent Device.

k) Grounded Conductor

Page 30, Article 100 Definitions., I General. --- A conductor connected to ground that, under normal conditions, may carry current. The white neutral conductor used in house wiring is a grounded conductor.

l)   Ground-Fault

Page 30, _Article 100 Definitions._, _I General._ --- An unintentional connection between a live conductor and earth.

m) Grounding Conductor

Page 30, _Article 100 Definitions._, _I General._ --- A conductor connected to ground that, under normal conditions, does not carry current. The green conductor used in house wiring is a Grounding Conductor.

n) Interrupting-Rating

Page 31, _Article 100 Definitions._, _I General._ --- At rated voltage the highest current that a device is designed to interrupt without failing or being damaged. It is not the device's overcurrent protection rating. For example, a fuse could have a 15 amperes overcurrent protection rating and a 10,000 amperes Interrupting-Rating.

o) Listed

Page 32, _Article 100 Definitions._, _I General._ --- A characteristic of equipment. Equipment that meets the standards of an organization may be put on their list of acceptable equipment. UL (Underwriters Laboratory) is an example of a listing organization. The electrical worker can assume that equipment listed by a reputable organization meets at least minimal standards. For example, a UL listed appliance would have a proper power cord.

p) Outlet

Page 32, _Article 100 Definitions._, _I General._ --- A point on the wiring system at which current is taken to supply utilization equipment.

q) Overcurrent

Page 32, _Article 100 Definitions._, _I General._ --- Current greater than the rated current of equipment or conductors. It can result from Overload, Short-Circuit, or Ground-Fault.

r) Overload

Page 32, _Article 100 Definitions._, _I General._ --- Operation of equipment or conductors at currents greater than their ratings for enough time to cause damage or dangerous overheating. Short-Circuits and Ground-Faults are not Overloads.

s) Panelboard

Page 33, _Article 100 Definitions._, _I General._ --- The panel that electrical control equipment is mounted on.

t) Qualified Person

Page 33, _Article 100 Definitions._, _I General._ --- Person who has skills and knowledge appropriate for the installation and operation of considered electrical equipment.

u) Raceway

Page 33, _Article 100 Definitions._, _I General._ --- An enclosed metallic or non-metallic channel for holding wires, cables, or busbars. Conduits and wireways are raceways.

v) Separately Derived System

Page 33, Article 100 Definitions., I General. --- An electrical source, other than a service, having no direct connections to circuit conductors of any other electrical source other than those established by grounding and bonding connections.

The following figures show examples of Separately Derived Systems and Non-Separately Derived Systems.

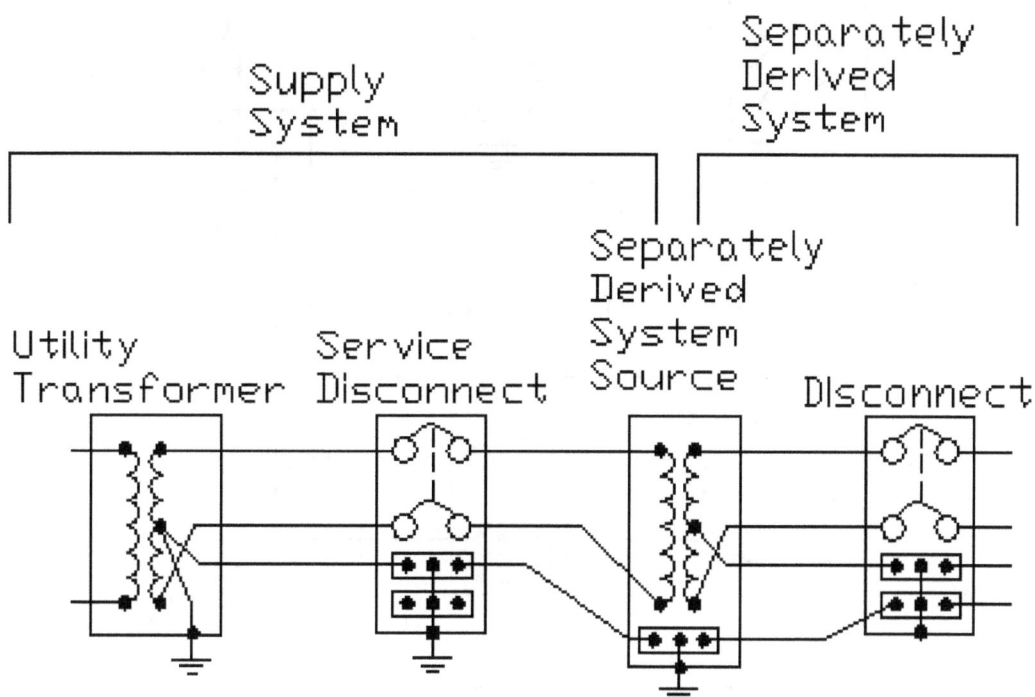

Figure 3-3   The output of this transformer is a Separately Derived System, since there is no direct connection to the Utility Supply System other than grounding. The NEC does not consider the transformer's magnetic input to output connection to be a direct connection.

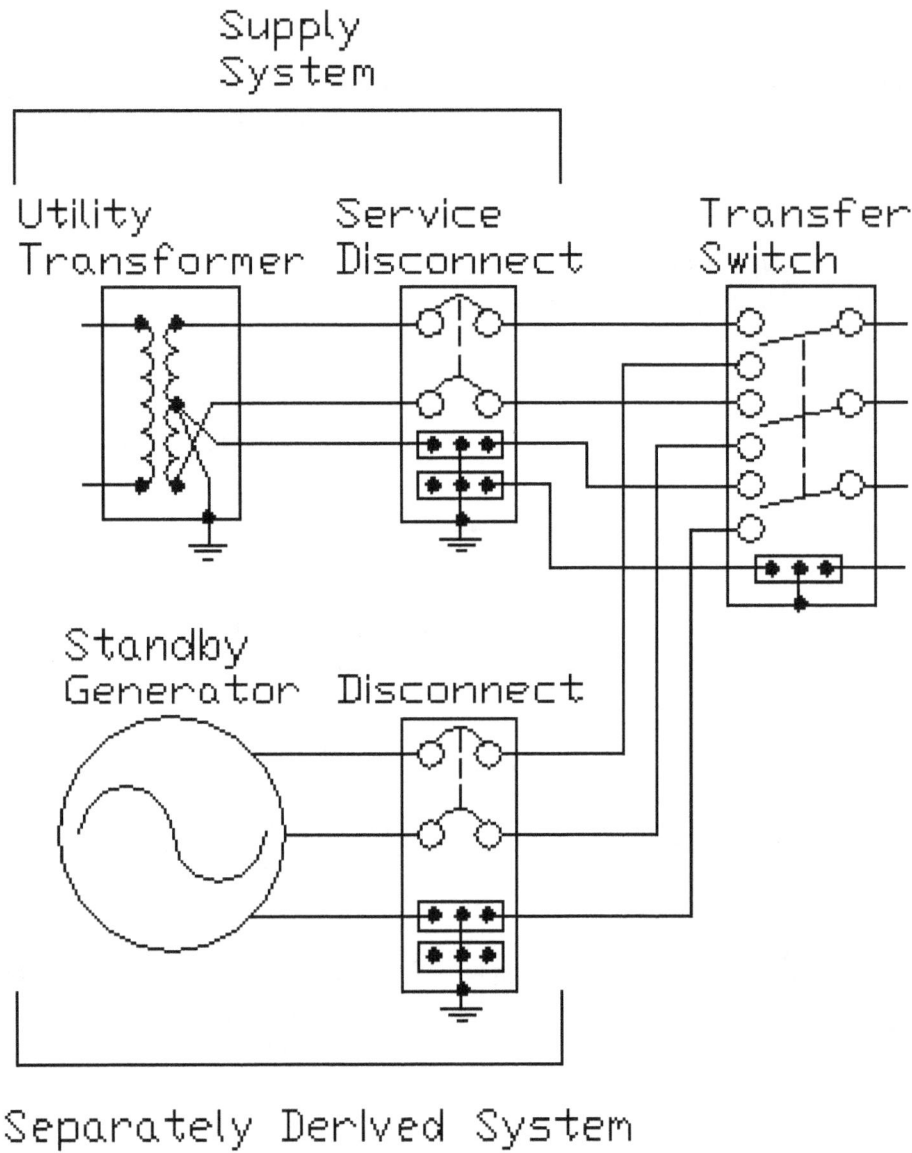

Figure 3-4 The output of this Standby Generator is a Separately Derived System since there is no direct connection to the Utility Supply System other than grounding. Notice how this uses a 3 pole transfer switch rather than the 2 pole transfer switch of the Non-Separately Derived System of Figure 3-5.

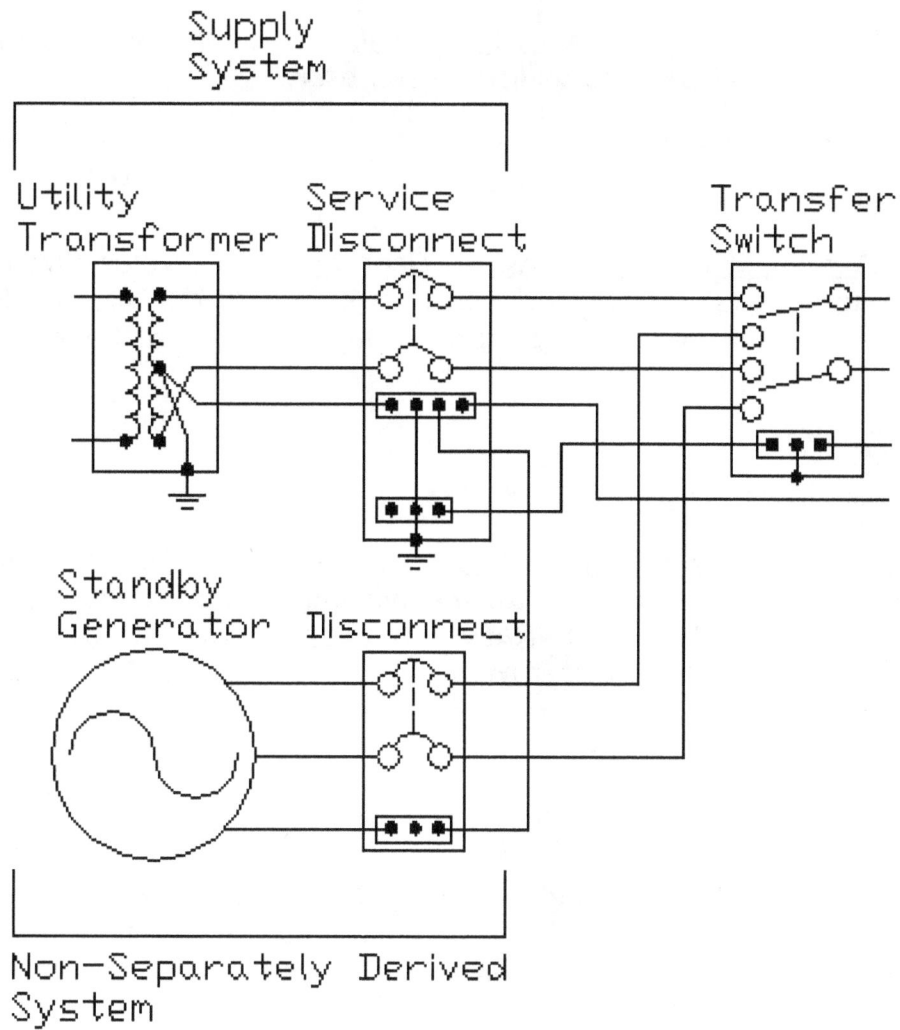

Figure 3-5 The output of this Standby Generator is a Non-Separately Derived System since there is a connection between the Standby Generator's neutral and that of the Utility Supply System.

w) Service

Page 33, Article 100 Definitions., I General. --- Cable and equipment for delivering electricity from the utility to the user.

x) Service Equipment

Page 34, Article 100 Definitions., I General. --- Circuit breaker(s) or switches(s) and fuse(s) at the load end of the Service Conductor that make up the main control and cutoff of the supply. In a typical house this would be the fuse panel or breaker panel.

y) Short-Circuit Current Rating

Page 34, Article 100 Definitions., I General. --- The Short-Circuit current that may be passed through a device without harming it. Note: This is different from the Interrupting-Rating because the device does not have to interrupt the current. Short-Circuit Current Rating is greater than Interrupting-Rating.

z) Ungrounded

Page 35, Article 100 Definitions., I General. --- Not connected to ground or connected to a conductive body that connects to ground. Ungrounded conductors are usually electrically charged relative to ground.

6) What is the minimum time for a *NEC* "continuous load"?

Page 29, Article 100 Definitions., I General., Continuous Load. --- Three hours.

7) What is an intersystem bonding termination device used for?

Page 31, Article 100 Definitions., I General. and *Page 123, Section 250.94 Bonding for Other Systems.* --- Connecting the grounds of a communication system to the main grounding electrode system.

8) What is the minimum Interrupting-Rating for Service, Feeder, and Branch Overcurrent Devices?

Page 32, _Article 100 Definitions_, _I General._, _Overcurrent Protective Device, Branch-Circuit._ --- The minimum Interrupting-Rating is 5,000 amperes. Note: Interrupting-Rating is the maximum current where a device can operate without being damaged. On an Overcurrent Device the Interrupting-Rating is always greater that the rated overcurrent value. For example, a common 120 volts 20 amperes circuit breaker would have at least a 5,000 amperes Interrupting-Rating.

9) Does the _NEC_ require that listed or labeled equipment be installed and used in accordance with instructions included in the listing or labeling?

Page 37, _Section 110.3 Examination_, _Identification, Installation, and Use of Equipment._, _(B) Installation and Use._ --- Yes.

10) What words does the _NEC_ use to state that electrical equipment and wiring be neatly and conventionally installed?

Page 37, _Section 110.12 Mechanical Execution of Work._ --- The _NEC_ states that, "Electrical equipment shall be installed in a neat and workmanlike manner." The words "neat and workmanlike" are used in many Articles.

11) May unused openings to electrical equipment be left open?

Page 37, _Section 110.12 Mechanical Execution of Work._, _(A) Unused Openings._ --- No, in most cases, they should be covered.

12) May conductors of dissimilar metals be intermixed in a terminal where the metals have contact with each other?

Page 38, _Section 110.14 Electrical Connections._ --- No.

13) Does the *NEC* allow house wiring to be extended by simply stripping, joining by twisting, and insulating with tape additional lengths of wire? (Information is available from several Sections.)

<u>Page 38</u>, <u>Section 110.14 Electrical Connections.</u>, <u>(B) Splices.</u> --- This indicates that simply twisting conductors together is not sufficient.

<u>Page 44</u>, <u>Section 110.31 Enclosure for Electrical Installations.</u>, <u>(B) Indoor Installations.</u>, <u>(1) In Places Accessible to Unqualified Persons.</u> --- This indicates that a junction box should be used over the splice point, if an unqualified person may come in contact with the splice.

<u>Page 148</u>, <u>Section 300.13 Mechanical and Electrical Continuity — Conductors.</u>, <u>(A) General.</u> --- This states that generally conductors shall be continuous without splices.

14) It is often difficult for an electrical inspector to prove that an electrical installation was not made in a "neat and workmanlike manner". However, it is easier to find other code violations. Cite a few common code violations. (Information is available in several Sections.)

<u>Page 38</u>, <u>Section 110.14 Electrical Connections.</u>, <u>(A) Terminals.</u> --- Terminal parts shall ensure a good connection (they may not be loose).

<u>Page 54</u>, <u>Section 210.8 Ground-Fault Circuit-Interrupter Protection for Personnel.</u>, <u>(A) Dwelling Units.</u> --- In dwellings, Ground-Fault Circuit-Interrupter (GFCI) should be installed in bathrooms, kitchens, and other locations.

<u>Page 94</u>, <u>Section 240.4 Protection of Conductors.</u> --- Conductors, other than flexible cords, flexible cables, and fixture wires, shall be protected against overcurrent in accordance with their ampacities, as specified on <u>Page 156</u>, <u>Section 310.15</u>, <u>Ampacities for Conductors Rated 0–2000 Volts.</u>

<u>Page 148</u>, <u>Section 300.11 Securing and Supporting.</u>, <u>(A) Secured in Place.</u> --- Raceways, cable assemblies, etc. must be securely fastened in place.

<u>Page 148</u>, <u>Section 300.14 Length of Free Conductors at Outlets.</u>, <u>Junctions, and Switch Points.</u> --- The length of free conductors at outlets should be at least 150 mm (6 in.).

15) Does the *NEC* allow conductors to have higher temperature ratings than their terminations?

<u>Page 38</u>, <u>Section 110.14 Electrical Connections.</u>, <u>(C) Temperature Limitations.</u> --- Yes. The ampacity of the conductors shall be selected so as to not exceed the maximum temperature ratings of their terminations. However, higher temperature rating conductors may be used provided there are "ampacity adjustments" (deratings) so that they do not exceed their terminations' maximum temperature ratings.

16) Where should Arc-Flash warnings be posted?

<u>Page 39</u>, <u>Section 110.16 Arc-Flash Hazard Warning.</u> --- On electrical equipment, like switchboards, that are likely to require examination or maintenance.

17) A 480 volts ac safety switch box is to be installed in a corridor, as shown in the following drawing. The walls are concrete. What are the minimum dimensions allowed by the *NEC*? (Information is required from a Section and Table.)

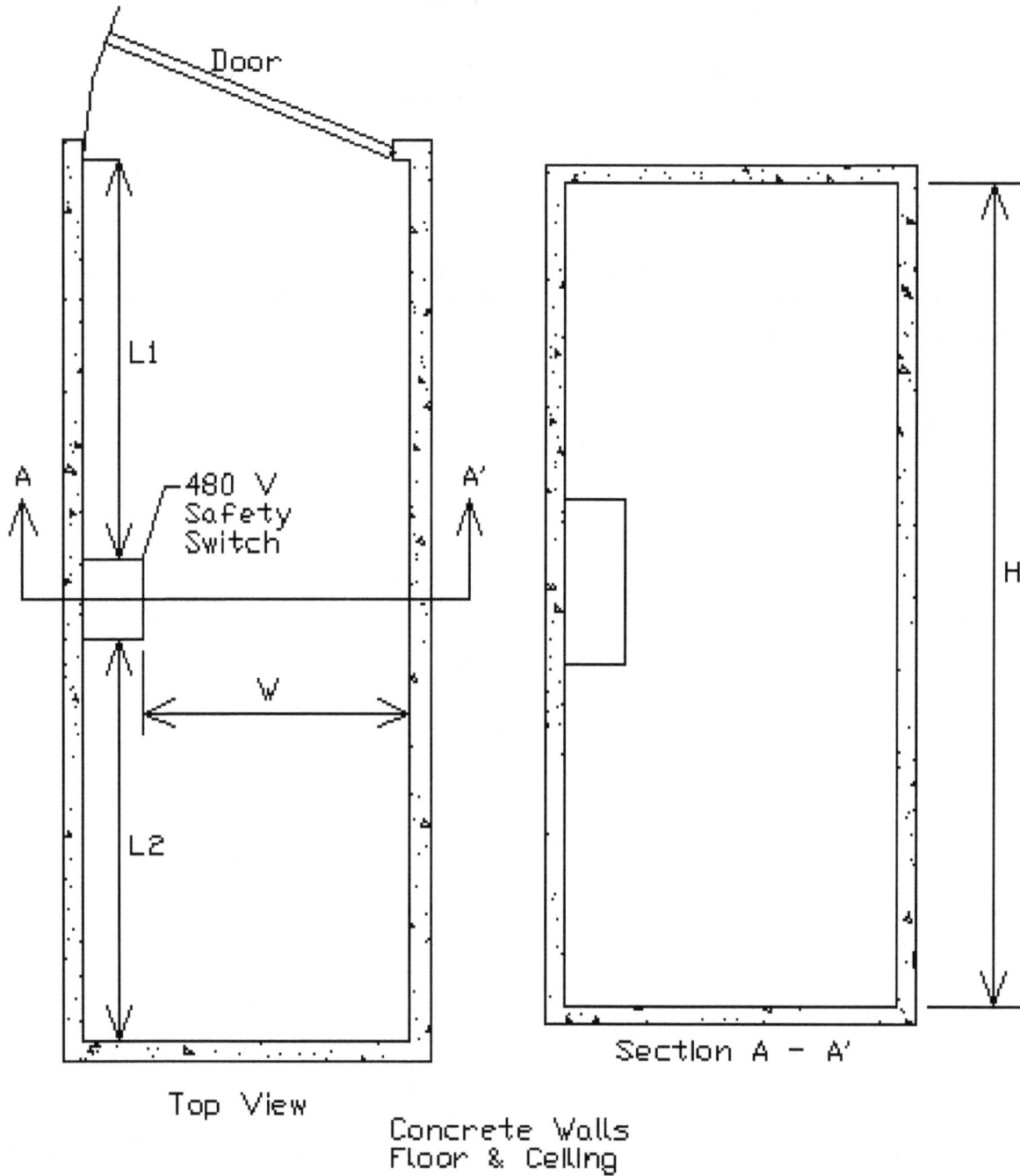

Figure 3-6 (same as Figure 2-3 of Section 2.0) 480 volts safety switch working space.

Page 40, _Table 110.26(A)(1) Working Spaces._, _Condition 2._ --- Concrete shall be considered as grounded.

Page 40, _Table 110.26(A)(1) Working Spaces._, _Condition 2._ --- Minimum clearance distance, W, is 1.07 m (3 ft 6 in.).

Page 40, _Section 110.26 Spaces About Electrical Equipment._, _(A) Working Space._, _(2) Width of Working Space._ --- The width of the working space, L1 +width of the safety switch + L2, needs to be at least 762 mm (30 in.).

Page 40, _Section 110.26 Spaces About Electrical Equipment._, _(A) Working Space._, _(3) Height of Working Space._ --- The height of the working space, H, needs to be at least 2 m (6 ft 6 in.).

18) Would the _NEC_ allow the storage of boxes under the safety switch of Figure 2-3?

Page 41, _Section 110.26 Spaces About Electrical Equipment._, _(B) Clear Spaces._ --- No.

19) When is an electrical room required to have two exits?

Page 41, _Section 110.26 Spaces About Electrical Equipment._, _(C) Entrance to and Egress from Working Space._, _(2) Large Equipment._ --- Equipment rated over 1200 amperes or more than 1.8 m (6 ft) wide containing overcurrent, switching, or control devices shall be in a room that has at least one entrance and one egress.

20) According to the _NEC_, when must live parts be guarded against accidental contact by the use of approved enclosures or by other _NEC_ approved means?

Page 42, _Section 110.27 Guarding of Live Parts._, _(A) Live Parts Guarded Against Accidental Contact._ --- The _NEC_ states that live parts operating at 50 volts or more shall be guarded against accidental contact.

21) What are the different indoor and outdoor enclosure types?

Page 43, _Table 110.28 Enclosure Selection._ --- This table gives enclosure types and enclosure capabilities.

22) Are equipment grounding conductors required to be bare or insulated?

Page 46, Section 110.54 Bonding and Equipment Grounding Conductors., (B) Equipment Grounding Conductors. --- Both insulated and bare are permitted.

23) May a grounded conductor (neutral conductor) depend on a metal enclosure or metal cable armor for continuity?

Page 49, Section 200.2 General., (B) Continuity. --- No.

24) May neutral conductors be used for more than one branch-circuit?

Page 49, Section 200.4 Neutral Conductors., (A) Installation. --- No.

25) What conductors use white insulation?

Page 50, Section 200.7 Use of Insulation of a White or Gray Color or with Three Continuous White or Gray Stripes., (A) General. --- White insulation shall only be used for grounded conductors (neutral conductors).

26) What are the current ratings of multi-outlet branch-circuits?

Page 51, Section 210.3 Rating <of Branch-Circuits>. --- 15, 20, 30, 40, and 50 amperes.

27) What Specific–Purpose Branch-Circuits are described in the *NEC*?

Page 52, Table 210.2 Specific-Purpose Branch-Circuits. --- This table gives Article and Section numbers for specific branch-circuits from Air-conditioning to X-ray equipment.

28) How do you find the Overcurrent Device ratings and conductor ratings needed for X-ray equipment? (Information can be found in a Table and Section)

Page 52, Table 210.2 Specific-Purpose Branch-Circuits. --- Table 210.2 refers to Page 595, Section 660.6, Rating of Supply Conductors and Overcurrent Protection.

29) What colors does the *NEC* specify for dc ungrounded conductors smaller than 6 AWG?

Page 53, Section 210.5 Identification for Branch-Circuits., Identification of Ungrounded Conductors., (2) Branch-Circuits Supplied From Direct-Current Systems. --- Positive conductors may have red insulation, insulation that is not black, white, grey, or green but has a continuous red stripe on it, or insulation that is not black, white, grey, or green but has a +, POS, or POSITIVE printed on it. Negative conductors may have black insulation, insulation that is not red, white, grey, or green but has a continuous black stripe on it, or insulation that is not red, white, grey, or green but has a - , NEG or NEGATIVE printed on it.

30) Where are GFCIs (Ground-Fault Circuit Interrupters) required?

Page 54, Section 210.8 Ground-Fault Circuit-Interrupter Protection for Personnel. --- They are now required in bathrooms, garages, outdoor areas, crawl spaces, unfinished basements, kitchens, sinks, boathouses, bathtubs and showers, and laundry areas. Exceptions may exist for some of these locations.

31) Where are AFCIs (Arc Fault Circuit Interrupters) required?

Page 56, Section 210.12 Arc-Fault Circuit-Interrupter Protection. --- They are now required in all new dwelling units, new modification of dwelling units, and dormitories. Note: Some Authorities Having Jurisdiction, such as local governments, do not require AFCIs.

32) What is the minimum ampacity of a branch-circuit conductor? (Information is required from several Sections.)

Page 57, Section 210.19 Conductors — Minimum Ampacity and Size., (A) Branch-Circuits Not More Than 600 Volts., (1) General. --- Not less than the maximum load.

Page 58, Section 210.19 Conductors — Minimum Ampacity and Size., (B) Branch-Circuits Over 600 Volts., (1) General. --- Not less than 125% of the load of equipment that will be operated simultaneously.

33) Does the *NEC* specify the percent voltage drop from the service to the load for circuits that have voltages not more than 600 volts? (Information is available from several Sections.)

Page 57, *Section 210.19 Conductors — Minimum Ampacity and Size.*, *(A) Branch-Circuits Not More Than 600 Volts.*, *Informational Note No. 4:.* --- No, it does not specify a voltage drop. However, *Informational Note No. 4* states that reasonable efficiency of operation will be obtained with a branch drop not exceeding 3% and combined feeder and branch drop not exceeding 5%. The same is stated on *Page 64*, *Section 215.2 Minimum Rating and Size.*, *(A) Feeders Not More Than 600 Volts.*, *Informational Note No. 2:.*

34) A branch-circuit supplies a continuous load of 10 amperes and an intermittent load of 15 amperes. What Overcurrent Device protection rating should be selected? If THHN copper conductor is to be used at an ambient temperature of 30°C what gauge should be selected? (Information is required from a Section and several Tables.)

Page 58, *Section 210.20 Overcurrent Protection.*, *(A) Continuous and Noncontinuous Loads.* --- 15 + 1.25 x 10 = 27.5 amperes. A 30 amperes circuit breaker should be selected.

Page 161, *Table 310.15(B)(16) (formerly Table 310.16) Allowable Ampacities of Insulated Conductors Rated Up to and Including 2000 Volts, 60°C Through 90°C (140°F Through 194°F), Not More Than Three Current-Carrying Conductors in Raceway, Cable, or Earth (Directly Buried), Based on Ambient Temperature of 30°C (86°F)*. --- 12 AWG would be protected by the selected 30 amperes breaker.

However, some Articles of the *NEC* would not accept 12 AWG for a 30 amperes circuit. *Page 60*, *Table 210.24 Summary of Branch-Circuit Requirements.* --- This indicates that the next larger gauge, 10 AWG, should be used.

35) What receptacle ampere ratings are required in single receptacles?

Page 58, *Section 210.21 Outlet Devices.*, *(B) Receptacles. (1) Single Receptacle on an Individual Branch Circuit.* --- Receptacles must be rated to at least their branch-circuit ampere rating.

36) What are standard current ratings for receptacles in multi-receptacle branch-circuits?

Page 58, Section 210.21 Outlet Devices., (B) Receptacles., (3) Receptacle Ratings. --- 15, 15 or 20, 30, 40 or 50, and 50 amperes.

37) How many small appliance receptacles are required for a kitchen countertop?

Page 61, Section 210.52 Dwelling Unit Receptacle Outlets., (C) Countertops., (1) Wall Countertop Spaces. --- There should be enough receptacles so that no point along the wall side of the counter is more than 600 mm (24 in.) from a receptacle.

38) What Section provides general and specific information on feeder circuit cable ampacities?

Page 64, Section 215.2 Minimum Rating and Size.

39) Is feeder Ground-Fault Protection required?
a) If line to line voltage is 480 volts, it is an ungrounded Δ system, and the disconnect is rated for 1500 amperes.
b) If line to line voltage is 480 volts, it is a solidly grounded Y system, and the disconnect is rated for 1500 amperes.
c) If line to line voltage is 208 volts, it is a solidly grounded Y system, and the disconnect is rated for 900 amperes.

Page 65, Section 215.10 Ground-Fault Protection of Equipment. --- a) No, this is not a solidly grounded Y system. b) Yes. c) No, the disconnect's current rating is less than 1000 amperes.

40) A school has a floor space of about 4,924 m^2 (53,000 ft^2). What would be the expected lighting load in kilovolt-amperes?

Page 68, Table 220.12 General Lighting Loads by Occupancy. --- In the table the estimate for lighting loads in a school classroom is 33 volt-amperes/m². From this the estimated lighting load is 33 x 4,924 = 162,492 volt-amperes = 162 kilovolt-amperes.

41) What Article covers outside branch and feeder circuits?

Page 76, Article 225 Outside Branch-Circuits and Feeders.

42) May cable carrying 1200 volts be installed outside and above ground in flexible conduit? (Information is available in several Sections.)

Page 77, Section 225.10 Wiring on Buildings (or Other Structures). and Page 153, Section 300.37 Aboveground Wiring Methods. --- No, outside this cable should be installed in rigid conduit.

43) Where is it stated that a means of disconnecting a service is required where ungrounded conductors enter a building?

Page 80, Section 225.31 Disconnecting Means. --- On a house this is in the service panel.

44) How many Service Disconnects are allowed in one location? (Information is available from several Sections.)

Page 80, Section 225.33 Maximum Number of Disconnects., (A) General. and Page 90, Section 230.71 Maximum Number of Disconnects., (A) General. --- There is a maximum of six.

45) A building that has more than one service must have a plaque placed at each service denoting all other services. Where is this stated in the *NEC*? (Information is available from several Sections.)

Page 80, Section 225.37 Identification. and Page 84, Section 230.2 Number of Services., (E) Identification.

46) What *NEC* figure graphically shows the relative locations of the Serving Utility, Service-Entrance Conductors, Service Equipment, and Branch-Circuit Feeders?

Page 83, Section 230.1 Scope., Figure 230.2 Services.

47) What *NEC* requirements apply to a building that receives electrical power from multiple sources?

Page 83, *Section 230.2 Number of Services.*

48) Usually a building may have only one service (supply). What are some exceptions where more than one supply is allowed?

Page 83, *Section 230.2 Number of Services.*, *parts (A) to (D).* --- Fire pumps, emergency systems, legally required standby systems, optional standby systems, parallel power production systems, systems designed for multiple sources to enhance reliability, multi-occupancy buildings where there is no space for service equipment, very large buildings, large power capacity requirements, and different voltage, frequency or phase requirements.

49) If a Service Conductor is to have a continuous load of 100 amperes and a noncontinuous load of 50 amperes what is its minimum ampacity?

Page 87, *230.42 Minimum Size and Rating.*, *(A) General.*, *(1).* --- The *NEC* gives three ways of calculating this. Method (1) determines the minimal ampacity is 1.25 x 100 + 50 = 175 amperes.

50) Where does the *NEC* state that Service-Entrance Conductors need to be protected against physical damage?

Page 88, *Section 230.50 Protection Against Physical Damage.*

51) What is a Service Disconnect?

Page 89, *Section 230.70 General.* --- Although the *NEC* often uses the words "Service Disconnect" and "Service Disconnecting Means", it does not clearly define them. A Service Disconnect is the first accessible point where a building owner can manually operate a switch that disconnects the building from the Utility Service Entrance electrical conductors. The Service Disconnect also provides protection against overcurrent. In new residential construction these are usually circuit breaker(s) called "Main Circuit Breaker(s)" and are located in the Service Panel.

52) Does the *NEC* allow the home owner to install a fuse box in his shower?

Page 89, Section 230 VI. Service Equipment — Disconnecting Means, 230.70 General., (A) Location., (2) Bathrooms. --- No, it does not. This *NEC* Section states that "Service disconnecting means shall not be installed in bathrooms." A home's fuse box (or beaker panel) includes the service disconnect.

53) Is access to each occupant's own Service Disconnecting Means required in multiple-occupancy buildings?

Page 90, Section 230.72 Grouping of Disconnects., (C) Access to Occupants. --- Generally yes, however, if the Service Disconnect is under continuous building management supervision, it is not required.

54) If Service Conductors supply five Service Disconnect/Overload Devices, each set to 200 amperes, but the calculated load is 900 amperes, what is the minimum ampacity of the Service Conductors?

Page 91, 230.90 Where Required., (A) Ungrounded Conductor., Exception No. 3: --- At a minimum the ampacity of the Service Conductors must be the equal to the calculated load. Here, the 900 amperes calculated load is less than 5 x 200 = 1000 amperes of the Service Disconnect Overload Devices, so the minimum ampacity of the Service Conductors is 900 amperes.

55) What is a Tap Conductor?

Page 94, Section 240.2 Definitions., Tap Conductors. --- A Tap Conductor is a smaller conductor connected to (tapped off) a larger conductor. Tapped Conductors are short. An example circuit having two tapped conductors would be a 50 amperes Overcurrent Protective Device and conductor connected to two short 20 amperes conductors to a separately located range top and oven. However, the 20 amperes tap conductors are not protected from 20 amperes overcurrent by their own 20 amperes Overcurrent Devices. Their only overcurrent protection would be the 50 amperes Overcurrent Protective Device that protects the 50 amperes conductors.

56) Where does the *NEC* state that conductors shall be protected against overcurrent?

Page 94, Section 240.4 Protection of Conductors. --- "Conductors, other than flexible cords, flexible cables, and fixture wires, shall be protected against overcurrent in accordance with their ampacities..."

57) Do flexible cords, i.e. lamp cords, require overcurrent protection?

Page 94, Section 240.4 Protection of Conductors. --- No.

58) Does the *NEC* allow pennies to be used in place of fuses in a fuse box or circuit breakers to be taped to the ON position in a load panel?

Page 94, Section 240.4 Protection of Conductors. --- No, the NEC states that conductors shall be protected against overcurrent. A penny in place of a fuse and a taped to the ON position circuit breaker will not protect against overcurrent.

59) Are Overload and Short-Circuit Protection required on crane lifting electromagnets?

Page 94, Section 240.4 Protection of Conductors., (A) Power Loss Hazard. --- Overload protection is not required, as the overload of the crane circuit is not as dangerous as the crane suddenly dropping its load. However, Short-Circuit Protection is required.

60) When sizing overcurrent protection, sometimes the Overcurrent Protective Device current rating should be the next higher from the calculated current and sometimes it should be the next lower. Find examples of this in the *NEC*.

Page 94, Section 240.4 Protection of Conductors., (B) Overcurrent Devices Rated 800 Amperes or Less. --- Overcurrent Devices rated 800 amperes or less may be the next higher standard Overcurrent Device rating (above the ampacity of the conductors being protected) provided certain conditions are met.

Page 95, Section 240.4 Protection of Conductors., (C) Overcurrent Devices Rated over 800 Amperes. --- Overcurrent Devices rated more than 800 amperes need to be rated to the next lower standard Overcurrent Device rating (below the ampacity of the conductors being protected).

61) What are the maximum Overcurrent Device ratings for 18, 16, 14, 12, and 10 AWG copper wire?

Page 95, Section 240.4 Protection of Conductors., (D) Small Conductors., (1) 18 AWG Copper., (2) 16 AWG Copper., (3) 14 AWG Copper., (5) 12 AWG Copper., (7) 10 AWG Copper. --- Generally, the device ratings should be 7, 10, 15, 20, and 30 amperes.

62) When does the *NEC* consider extension cords protected?

Page 96, 240.5 Protection of Flexible Cords, Flexible Cables, and Fixture Wires., (B) Branch-Circuit Overcurrent Device., (3) Extension Cord Sets. --- The *NEC* considers them protected when they are used within their extension cord listing range.

63) What are some standard fuse and circuit breaker ratings?

Page 96, Section 240.6 Standard Ampere Ratings., (A) Fuses and Fixed-Trip Circuit Breakers. --- Some of the standard ratings are 15, 20, 25, 30, 35, 40, 45, and 50 amperes.

64) If an adjustable setting circuit breaker does not have restricted access to its adjustment how must it be used?

Page 97, Section 240.6 Standard Ampere Ratings., (B) Adjustable-Trip Circuit Breakers. --- If the access cannot be restricted from unqualified persons then only the circuit breaker's maximum setting may be used.

65) May ordinary single fuses be connected in parallel?

Page 97, Section 240.8 Fuses or Circuit Breakers in Parallel. --- No, they may not. However, fuses may be connected in parallel if they were factory-assembled that way as a unit.

66) What conductors shall have Overcurrent Protection?

Page 97, Section 240.15 Ungrounded Conductors., (A) Overcurrent Device Required. --- Each Ungrounded Conductor shall be protected by a fuse or circuit breaker.

67) Where should a conductor's Overcurrent Protection be located?

Page 98, Section 240.21 Location in Circuit. --- Generally, it shall be located at the point the conductor receives its supply.

68) What is a major difference between the 0 to 3 m (0 to 10 ft) and >3 to 7.5 m (>10 to 25 ft) feeder tap rules?

Page 98, Section 240.21 Location in Circuit., (B) Feeder Taps., (1) Taps Not over 3 m (10 ft) Long. and (2) Taps Not over 7.5 m (25 ft) Long. --- A major difference is that a 0 to 3 m (0 to 10 ft) feeder tap may use tap conductors that have a minimum ampacity of 1/10 of its feeder's Overcurrent Protection while a >3 to 7.5 m (>10 to 25 ft) tap needs to use larger tap conductors that have a minimum capacity of 1/3 of its feeder's Overcurrent Protection.

69) Is it necessary to place Overcurrent Protection on a transformer's secondary? (Information is required from several Sections.)

Page 99, Section 240.21 Location in Circuit., (C) Transformer Secondary Conductors., (1) Protection by Primary Overcurrent Device. and Page 361, Section 450.3 Overcurrent Protection. --- Usually, but not always, in some cases the "tap rules" may be applied and only primary Overcurrent Protection is sufficient.

70) When does the *NEC* require arc energy reduction?

Page 103, 240.87 Arc Energy Reduction. --- The *NEC* requires actions if an overcurrent trip device is set to 1200 amperes or higher.

71) What is the Article on grounding?

Page 106, Article 250 Grounding and Bonding.

72) What are the reasons for grounding?

Page 108, Section 250.4 General Requirements for Grounding and Bonding., (A) Grounded Systems., (1) Electrical System Grounding. --- To limit voltages from lightning, surges, or unintended connections with high voltages and to stabilize normal operating voltages.

73) What is an effective Ground-Fault current path?

Page 108, Section 250.4 General Requirements for Grounding and Bonding., (A) Grounded Systems., (5) Effective Ground-Fault Current Path. --- A low impedance circuit that can trip an Overcurrent Device or facilitate the operation of a Ground-Fault Detector. It should be capable of safely carrying likely maximum ground-fault current.

74) May an outdoor receptacle be grounded locally through its own ground rod?

Page 108, Section 250.4 General Requirements for Grounding and Bonding., (A) Grounded Systems., (5) Effective Ground-Fault Current Path. and (B) Ungrounded Systems., (4) Path for Fault Current. --- No, the earth is not considered an effective fault-current path. Grounding conductors connect back to the service.

75) May bonding and grounding connections be made with solder?

Page 109, Section 250.8 Connection of Grounding and Bonding Equipment., (B) Methods Not Permitted. --- Grounding and bonding connections that depend solely on solder are not allowed.

76) Where does the *NEC* state that a 120/240 volts ac system, such as that used in a house, shall be grounded?

Page 109, Section 250.20 Alternating-Current Systems to Be Grounded., (B) Alternating-Current Systems of 50 Volts to 1000 Volts.

77) What markings must be put on a 50 to 1000 volts ungrounded ac system?

Page 110, Section 250.21 Alternating-Current Systems of 50 Volts to 1000 Volts Not Required to Be Grounded., (C) Marking. --- "Caution: Ungrounded System Operating — ____ Volts Between Conductors"

78) What circuits should not be grounded?

Page 110, Section 250.22 Circuits Not to Be Grounded. --- Some examples are cranes over combustible fibers, health care facilities, electrolytic cells, and low voltage lighting.

79) May a grounded conductor (neutral conductor) be reconnected to ground at its load equipment?

Page 110, Section 250.24 Grounding Service-Supplied Alternating-Current Systems., (A) System Grounding Connections., (5) Load-Side Grounding Connections. --- No, the grounded conductor (neutral conductor) was already connected to ground at the service. It shall not be reconnected to ground at the load or any other place.

80) 3/0 AWG copper wire is used for the ungrounded Service Conductors on a 240/120 volts ac system. The Service Conductors are in a single conduit. What is the minimum size for a copper wire grounded Service Conductor? (Information is available from a Section and Table.)

Page 111 , Section 250.24 Grounding Service-Supplied Alternating-Current Systems., (C) Grounded Conductor Brought to Service Equipment., (1) Sizing for a Single Raceway. and Page 124, Table 250.102(C)(1) Grounded Conductor, Main Bonding Jumper, System Bonding Jumper, and Supply-Side Bonding. --- From the table the minimum size is 4 AWG copper wire.

81) In the Separately Derived System of Figure 2-2 the transformer of the Separately Derived System Source does not have its output mid-tap connected directly to its metal cabinet. This is to avoid creating a parallel conducting path ground loop.
a) Where does the *NEC* state that parallel paths should be avoided?
b) If the ungrounded conductor is 3/0 AWG copper wire what is the minimum size for the system bonding jumper and the supply-side bonding jumper?
(Information is required from a Section and Table.)

a) <u>Page 112, Section 250.30 Grounding Separately Derived Alternating-Current Systems., (A) Grounded Systems., (1) System Bonding Jumper., Exception No. 2</u>.

b) <u>Page 124, Table 250.102(C)(1) Grounded Conductor, Main Bonding Jumper, System Bonding Jumper, and Supply-Side Bonding Jumper for Alternating-Current Systems.</u> --- 4 AWG copper wire could be used for each.

82) If parallel ungrounded conductors are run in parallel raceways in a Separately Derived System, are parallel grounded conductors required in each raceway?

<u>Page 112, Section 250.30 Grounding Separately Derived Alternating-Current Systems., (A) Grounded Systems., (3) Grounded Conductor., (b) Parallel Conductors in Two or More Raceways.</u> --- Yes.

83) May a common tapped grounding electrode conductor be used with multiple Separately Derived Systems?

<u>Page 113, Section 250.30 Grounding Separately Derived Alternating-Current Systems., (A) Grounded Systems., (6) Grounding Electrode Conductor, Multiple Separately Derived Systems</u>. --- Yes.

84) What is the main grounding rule for two or more buildings supplied by feeders or branches of the same system?

<u>Page 114, Section 250.32 Buildings or Structures Supplied by a Feeder(s) or Branch-Circuit(s).</u> --- Connect to grounding electrodes at each building.

85) Is a portable standby generator's frame required to be grounded?

Page 115, Section 250.34 Portable and Vehicle-Mounted Generators., (A) Portable Generators. --- Generally, no.

86) When is a high impedance neutral to ground allowed?

Page 116, Section 250.36 High-Impedance Grounded Neutral Systems. --- On 3-phase 480 to 1000 volts systems where only qualified personnel will do servicing, ground detectors are used, and line to neutral loads are not used.

87) Should electrodes be bonded together to form a grounding electrode system? (Information is required from several Sections.)

Page 117, Section 250.50 Grounding Electrode System. --- Yes, all available electrodes should be bonded together to form a grounding electrode system. This includes underground metal water pipes, the metal building frame, and the other grounding electrodes mentioned in Page 117, Section 250.52 Grounding Electrodes., (A) Electrodes Permitted for Grounding.

88) What length of a metal water pipe must be in contact with the earth for it to be used a grounding electrode?

Page 117, Section 250.52 Grounding Electrodes., (A) Electrodes Permitted for Grounding., (1) Metal Underground Water Pipe. --- 3 m (10 ft) or more.

89) Are gas pipes and aluminum permitted to be grounding electrodes?

Page 117, Section 250.52 Grounding Electrodes., (B) Not Permitted for Use as Grounding Electrodes. --- No.

90) What is the maximum resistance of a single rod, pipe, or plate grounding electrode to ground before a supplemental electrode is required?

Page 118, Section 250.53 Grounding Electrode System Installation., (A) Rod, Pipe, and Plate Electrodes., (2) Supplemental Electrode Required. --- 25 ohms.

91) A grounding electrode shall have at least 2.44 m (8 ft) of its length in contact with the soil. What are the alternatives to driving the grounding electrode straight down?

Page 118, Section 250.53 Grounding Electrode System Installation., (G) Rod and Pipe Electrodes. --- The electrode can be driven at an angle of up to 45 degrees or can be buried horizontally at least 750 mm (30 in.) deep.

92) Are splices allowed on grounding electrode conductors?

Page 119, Section 250.64 Grounding Electrode Conductor Installation., (C) Continuous. --- Generally no, unless the splicing is done with irreversible compression connectors, exothermic welding, or other permanent methods.

93) If ungrounded service entrance conductors are 1/0 AWG copper, what is the minimum size of the copper grounding electrode conductor?

Page 121, Table 250.66 Grounding Electrode Conductor for Alternating-Current Systems. --- 6 AWG.

94) By what methods may a grounding electrode conductor be connected to its grounding electrode?

Page 121, Section 250.70 Methods of Grounding and Bonding Conductor Connection to Electrodes. --- With exothermic welding, listed lugs, listed clamps, or listed pressure connectors.

95) What must be the current capacity of a bonding jumper?

Page 122, V. Bonding., Section 250.90 General. --- It shall be able to safely conduct any fault current likely to be imposed.

96) Are standard conduit lock nuts sufficient for bonding?

Page 122, Section 250.92 Services., (B) Method of Bonding at the Service. --- Standard lock nuts are acceptable for mechanical purposes, but for bonding a listed bonding jumper should be used.

97) What methods of bonding are acceptable for use in hazardous locations? (Information is required from several Sections.)

Page 124, Section 250.100 Bonding in Hazardous (Classified) Locations. and Page 122, Section 250.92 Services., (B) Method of Bonding at the Service., (2) to (4). --- Listed threaded and unthreaded couplings and other listed devices.

98) What may a bonding jumper be made of?

Page 124, Section 250.102 Bonding Conductors and Jumpers., (A) Material. --- Copper or another corrosion resistant material. It can be a wire, screw, bus, or similar conductor.

99) A branch-circuit is protected by a 30 amperes breaker. What should be the size of the aluminum bonding jumpers used in this load-side circuit? (Information is required from several Sections.)

Page 124, Section 250.102 Bonding Conductors and Jumpers., (D) Size — Equipment Bonding Jumper on Load Side of an Overcurrent Device. --- This should be found in Table 250.122.

Page 131, Table 250.122 Minimum Size Equipment Grounding Conductors for Grounding Raceway and Equipment. --- Aluminum bonding jumpers should be at least 8 AWG.

100) Ungrounded conductors in an ac service are 3/0 AWG copper. What is the minimum gauge of a copper Grounded Conductor, Main Bonding Jumper, System Bonding Jumper, and Supply-Side Bonding Jumper?

Page 124, Table 250.102(C)(1) Grounded Conductor, Main Bonding Jumper, System Bonding Jumper, and Supply-Side Bonding Jumper for Alternating-Current Systems. --- 4 AWG copper.

101) Should gas pipes and structural steel be bonded?

Page 125, Section 250.104 Bonding of Piping Systems and Exposed Structural Metal., (B) Other Metal Piping. --- Yes, if they are likely to become energized.

102) Is grounding required for a permanently located metal lathe that uses 460 volts ac 3-phase?

Page 126, Section 250.110 Equipment Fastened in Place (Fixed) or Connected by Permanent Wiring Methods. --- Yes, since it will be in contact with people, it may be damp (if cutting oil coolant is used), and it operates at over 150 volts to ground.

103) When should equipment metal parts be connected to an equipment grounding conductor? (Information is required from several Sections.)

Page 126, Section 250.110 Equipment Fastened in Place (Fixed) or Connected by Permanent Wiring Methods. and *Page 127, Section 250.112 Specific Equipment Fastened in Place (Fixed) or Connected by Permanent Wiring Methods.* --- Whenever exposed metal parts are close to energized conductors or are part of certain types of equipment.

104) Name some instances where equipment grounding is not required.

Page 127, Section 250.112 Specific Equipment Fastened in Place (Fixed) or Connected by Permanent Wiring Methods., (F) Garages, Theaters, and Motion Picture Studios. and *(I) Remote-Control, Signaling, and Fire Alarm Circuits.* --- Garages, theaters, and motion picture studios where voltages are less than 150 volts do not require equipment grounding. Some low voltage and low power remote-control, signaling, and fire-alarm circuits do not require equipment grounding.

105) What is the maximum length and current capacity of listed flexible metal conduit that is being used as a ground-fault current path?

Page 128, Section 250.118 Types of Equipment Grounding Conductors., (5). --- 1.8 m (6 ft) and 20 amperes.

106) Describe the coverings required for grounding conductors.

Page 129, Section 250.119 Identification of Equipment Grounding Conductors. --- The conductors may be bare or insulated. If insulated they should have either green insulation or green insulation with yellow stripes.

107) The color codes in common use for U.S. ac house wiring are:

Conductor Insulation Color Codes	
Color	**Function**
Black	Ungrounded conductor (hot wire)
Red	Ungrounded conductor (hot wire)
Blue	Ungrounded conductor (hot wire)
White coded black[*]	Ungrounded conductor (hot wire)[*]
White	Grounded conductor (neutral wire)
Green	Grounding conductor
Green & yellow stripe	Grounding conductor
Bare copper	Grounding conductor
[*]Usually white wires are grounded. However, occasionally a white wire used in a switch loop will need to be ungrounded. In that case it is marked with a black tape or a spot of black paint at its ends.	

Are these color codes in the *NEC*? (Information is required from several Sections.)

Only some, the ones for grounding conductors, are in the NEC. See Page 129, Section 250.119 Identification of Equipment Grounding Conductors. Also the color code for the grounded conductor (white) is in the NEC. See Page 50, Section 200.7 Use of Insulation of a White or Gray Color or with Three Continuous White or Gray Stripes. In various other places in the NEC, the black and red colors are only mentioned as colors that shall not be used on grounding conductors.

108) Conductors with green or green with yellow stripes insulation are usually equipment grounding conductors. Where could green and green and yellow stripe insulated conductors be used for other than equipment grounding?

Page 129, 250.119 Identification of Equipment Grounding Conductors., Exceptions 1 & 3: --- Some class 2 and 3 signaling cables under 50 volts, some special flexible cords, and some traffic signal cables may be green and not be equipment grounding conductors.

109) How close to the earth may aluminum equipment grounding conductors be terminated?

Page 129, 250.120 Equipment Grounding Conductor Installation., (B) Aluminum and Copper-Clad Aluminum Conductors. --- 450 mm (18 in.).

110) Are there cases where the equipment grounding conductor's gauge should be larger than that of the ungrounded conductors?

Page 130, 250.122 Size of Equipment Grounding Conductors., (A) General. --- No, the equipment grounding conductor is never required to be larger.

111) What gauge should the grounding conductor be in a small flexible cord?

Page 130, Section 250.122 Size of Equipment Grounding Conductors., (E) Flexible Cord and Fixture Wire. --- For ungrounded conductors 10 AWG or less the grounding conductor shall not be less than the ungrounded conductors and shall not be less than 18 AWG.

112) What is the minimum size of the copper grounding conductor for equipment protected by a 20 amperes circuit breaker?

Page 131, Table 250.122 Minimum Size Equipment Grounding Conductors for Grounding Raceway and Equipment. --- 12 AWG copper wire.

113) May the structural metal frame of a building be used as ac equipment grounding?

Page 132, Section 250.136 Equipment Considered Grounded., (A) Equipment Secured to Grounded Metal Supports. --- No.

114) When may a receptacle have an isolated ground, a ground that is connected to the grounding electrode conductor but not to its receptacle enclosure?

Page 133, Section 250.146 Connecting Receptacle Grounding Terminal to Box., (D) Isolated Ground Receptacles. --- Where it is necessary to reduce electrical noise on the grounding circuit.

115) What dc systems need to be grounded?

Page 134, Section 250.162 Direct-Current Circuits and Systems to Be Grounded., --- Two wire dc systems where the voltage is between 60 and 300 volts dc and three wire dc systems.

116) What is the minimum percent ampacity the neutral conductor may have compared to the phase conductors?

Page 136, Section 250.184 Solidly Grounded Neutral Systems., (A) Neutral Conductor., (2) Ampacity. --- The neutral conductor needs to have an ampacity sufficient to for the load it carries and not less than 33 1/3 percent of phase conductor ampacities.

117) What is the selection of a metal oxide surge arrestor based on?

Page 139, Section 280.4 Surge Arrestor Selection., Informational Note No. 2. --- The selection "...is based on consideration of maximum continuous operating voltage and the magnitude and duration of overvoltages..."

118) What are the "Wiring Methods and Materials" referred to in Article 300?

Page 142, Article 300 General Requirements for Wiring Methods and Materials. --- By "Wiring Methods and Materials", this Article refers to things that conductors go into and through, such as holes bored in wood, raceways, cable trays, splice boxes, etc.

119) The *NEC* says conductors of the same circuit should be kept together. Where is this stated in the *NEC*?

Page 142, Section 300.3 Conductors., (B) Conductors of the Same Circuit. --- Grouping conductors of the same circuit reduces magnetic effects on enclosures and overall circuit impedance.

120) May conductors rated to carry more than 1000 volts be next to those rated to carry less than 1000 volts?

Page 143, Section 300.3 Conductors., (C) Conductors of Different Systems., (2) Over 1000 Volts, Nominal. --- No, they need to go in separate wiring enclosures, cables, or raceways.

121) May active cables and conductors attached to ceiling joists be used as clotheslines? (Information is available in several Sections.)

Page 143, Section 300.4 Protection Against Physical Damage. --- No, using active cables and conductors as clotheslines would expose them to physical damage. Similar prohibitions also occur in NEC Sections on AC, MC and NM cables. See Page 193, Section 320.12 Uses Not Permitted.; Page 200, Section 330.12 Uses Not Permitted.; and Page 204, Section 334.15 Exposed Work., (B) Protection from Physical Damage.

122) If cables are to go through holes bored in joists, rafters, or wood members, how far back from the surface should the holes be?

Page 143, Section 300.4 Protection Against Physical Damage., (A) Cables and Raceways Through Wood Members., (1) Bored Holes. --- The holes should be bored so that the closest edge of them is 32 mm (1.25 in.) from the nearest edge of the wooden member. With this separation most nails will not be accidentally driven through the beam into the conductors.

123) How deep must conduit containing 120 volts ac wiring be buried beneath a concrete driveway?

Page 145, Table 300.5 Minimum Cover Requirements, 0 to 1000 Volts, Nominal, Burial in Millimeters (Inches). --- It should be at least 600 mm (24 in.) under the concrete.

124) What sort of conduit would be appropriate for underground use where the conduit may be subject to physical damage?

Page 146, Section 300.5 Underground Installations., (D) Protection from Damage., (4) Enclosure or Raceway Damage. --- Rigid metal conduit, intermediate metal conduit, RTRC-XW, Schedule 80 PVC conduit, or equivalent.

125) Are splice boxes required on buried cable splices? (Information is available from several Sections.)

Page 146, Section 300.5 Underground Installations., (E) Splices and Taps. and *Page 153, Section 300.50 Underground Installations., (D) Splices.* --- Splice boxes are not required provided the splices are watertight and made with suitable materials. Note: Underground there is no danger of fire or electric shock, so the *NEC* is able to reduce requirements there.

126) Does the *NEC* require stainless steel raceways to have a protective coating, as it does for other ferrous metal raceways?

Page 146, Section 300.6 Protection Against Corrosion and Deterioration., (A) Ferrous Metal Equipment., Exception: --- Stainless steel raceways are not required to have a protective coating.

127) May support wires be used as the sole support for hanging conduit from a ceiling?

Page 148, Section 300.11 Securing and Supporting, (A) Secured in Place. --- Generally no, conduit needs to be securely supported. Support wires, like those for hanging dropped ceilings are not secure enough.

128) What precaution should be taken when installing electrical conduit and cable to reduce the possibility of fire spread?

Page 151, Section 300.21 Spread of Fire or Products of Combustion. --- Cables and conduit should be installed with fire stops to impede fire spread.

129) May wiring be placed in ducts that carry dust, loose stock, or vapors?

Page 151, Section 300.22 Wiring in Ducts Not Used for Air Handling, Fabricated Ducts for Environmental Air, and Other Spaces for Environmental Air (Plenums)., (A) Ducts for Dust, Loose Stock, or Vapor Removal. --- *No, wiring should not go into such ducts.*

130) Where is there general information on conductor designations, insulations, markings, mechanical strengths, ampacity ratings, and uses?

Page 154, Article 310 Conductors for General Wiring.

131) May any insulated conductor in the *NEC* be used in dry locations?

Page 155, Section 310.10 Uses Permitted., (A) Dry Locations. --- *Yes.*

132) May conductors be used in ambient temperatures greater than their rated temperature?

Page 157, Section 310.15 Ampacities for Conductors Rated 0-2000 Volts., (A) General., (3) Temperature Limitation of Conductors. --- *No.*

133) What are the principal determinants of conductor temperature?

Page 157, Section 310.15 Ampacities for Conductors Rated 0-2000 Volts., (A) General., (3) Temperature Limitation of Conductors., Informational Note No. 1:. --- *Conductor temperature is determined by ambient temperature, resistance heating in the conductor, rate that heat dissipates from the conductor, and resistance heating of nearby conductors.*

134) If 30 conductors are contained in a conduit how much should the conductors' ampacities be reduced from that for 3 conductors in a conduit?

Page 160, Table 310.15(B)(3)(a) Adjustment Factors for More Than Three Current-Carrying Conductors. --- *This table shows that the ampacities should be reduced to 45% of the values determined in the 3-conductors-in-a-conduit ampacity tables.*

135) How many °C would the effective ambient temperature be increased on conduit that is mounted outside 380 mm (15 in.) above a roof?

Page 160, Table 310.15(B)(3)(c) Ambient Temperature Adjustment for Raceways or Cables Exposed to Sunlight on or Above Rooftops. --- It would be increased 14°C.

136) Two 12 AWG THHN copper conductors are to be used in conduit at an ambient temperature of 10°C. What is the maximum ambient temperature that these conductors may be used in? What is its current rating at its maximum ambient temperature? What is its current rating at 10°C? (Information is required from several Tables and a Section.)

Page 161, Table 310.15(B)(16) (formerly Table 310.16) Allowable Ampacities of Insulated Conductors Rated Up to and Including 2000 Volts, 60°C Through 90°C (140°F Through 194°F), Not More Than Three Current-Carrying Conductors in Raceway, Cable, or Earth (Directly Buried), Based on Ambient Temperature of 30°C (86°F)*. --- This gives the conductor's maximum ambient temperature as 90°C and it current capacity at 30°C as 30 amperes.

Page 158, Table 310.15(B)(2)(a) Ambient Temperature Correction Factors Based on 30°C (86°F). --- At 90°C the correction factor for these conductors is undefined.

Page 157, Section 310.15 Ampacities for Conductors Rated 0-2000 Volts., (B) Tables., (2) Ambient Temperature Correction Factors. --- The given equation determines that the conductors' current capacity at 90°C is 0 amperes.

Page 158, Table 310.15(B)(2)(a) Ambient Temperature Correction Factors Based on 30°C (86°F). --- At 10°C the correction factor for this conductor is 1.15, so the conductors' current capacity is then 1.15 x 30 = 34.5 amperes.

137) What *NEC* table gives the maximum number of conductors that may be stuffed into a metal box?

Page 186, Table 314.16(A) Metal Boxes.

138) What is the minimum length for a straight pull box connected to a Metric Designator 27 (Trade Size 1) EMT conduit?

Page 190, Section 314.28 Pull and Junction Boxes and Conduit Bodies., (A) Minimum Size., (1) Straight Pulls. --- The straight pull box should be at least eight times the Metric Designator (Trade Size) of the largest raceway. This is 8 x 27 = 216 mm (8 x 1 = 8 in.) long.

139) What are SE and USE cables?

Page 207, Section 338.2 Definitions. Service-Entrance Cable. --- They are above ground Service Entrance cables and Underground Service Entrance cables.

140) May type UF cable be used as underground service-entrance cable?

Page 208, Section 340.12 Uses Not Permitted. No, type UF cable shall not be used as service-entrance cable.

141) Where is PVC conduit not to be used?

Page 216, Section 352.12 Uses Not Permitted. --- Hazardous locations, support of luminaires, locations where physical damage is possible, in higher ambient temperatures, and theaters.

142) What is a wireway? (Information is required from two Sections.)

Page 237, Section 376.2 Definition. Metal Wireways. and Page 239, Section 378.2 Definition. Nonmetallic Wireways. --- A trough for housing and protecting electrical wires. Wires are laid into a wireway after the wireway is installed and a hinged or removable cover goes over it. A wireway is a raceway but a raceway is not necessarily a wireway.

143) What maximum percent fill is allowed for metal wireways?

Page 237, Section 376.22 Number of Conductors and Ampacity., (A) Cross-Sectional Areas of Wireway. --- The total of a wireway's conductor cross-sectional area (including insulation) may not exceed 20% of the cross-sectional area of the wireway.

144) Where do you find ampacities for flexible cords?

Page 269, Section 400.5 Ampacities for Flexible Cords and Cables.

145) May extension cords be used as a substitute for a structure's fixed wiring?

Page 271, Section 400.8 Uses Not Permitted. --- No.

146) Are splices allowed on flexible cords? (Information is required from a Section and Table.)

Page 271, Section 400.9 Splices. --- Generally no, however some hard service cords may be spliced. See *Pages 264* and *265, Trade Name* column in *Table 400.4 Flexible Cords and Cables.* for hard service cords.

147) A fixture is anything permanently attached to a building. A wall-mounted ac electric clock would be an example of an electrical fixture. An electrical fixture is powered by fixture wires. What are the largest and smallest wire gauges the *NEC* allows for fixture wires? (Information is required from a Table and Section.)

Page 277, Table 402.5 Allowable Ampacity for Fixture Wires. and *Page 277, Section 402.6 Minimum Size.* --- The largest mentioned is 10 AWG. The smallest allowed is 18 AWG.

148) Where are switches discussed in the *NEC*?

Page 277, Article 404 Switches.

149) How high may the operating handle of a switch or circuit breaker be above the floor?

Page 279, Section 404.8 Accessibility and Grouping., (A) Location. --- Not more than 2 m (6 ft 7 in.) above the floor.

150) How is an isolated ground receptacle identified?

Page 281, Section 406.3 Receptacle Rating and Type., (D) Isolated Ground Receptacles. --- An isolated ground receptacle is identified by an orange triangle on its face.

151) Where does the _NEC_ state that receptacles in damp and wet areas must have protection against moisture?

Page 284, Section 406.9 Receptacles in Damp or Wet Locations.

152) What is the maximum Overcurrent Device rating for a feeder to a 400 amperes panelboard? Where must the Overcurrent Device be mounted?

Page 288, Section 408.36 Overcurrent Protection. --- The maximum Overcurrent Device rating is the same as the panelboard, 400 amperes. It should be located on the panelboard or on its feeder input side.

153) Are open or bare incandescent bulb fixtures allowed in closets?

Page 293, 410.16 Luminaires in Clothes Closets., (B) Luminaire Types Not Permitted. --- No, if incandescent bulbs are used they must be surface or recessed mounted.

154) Is it necessary to leave open space around recessed lights, or may insulation be put next to them?

Page 298, Section 410.116 Clearance and Installation. --- Type IC Luminaires do not require space. Non-Type IC Luminaires require at least 13 mm (.5 in.) space on their sides and 75 mm (3 in.) space above.

155) Is ground fault protection required with a house's gutter de-icing cable?

Page 319, 426.28 Ground-Fault Protection of Equipment. --- Yes.

156) What is the *NEC* Article on motors? What table gives *NEC* Articles and Sections that have information on the application of electric motors in specific equipment? (Information is required from an Article and Table.)

Page 323, Article 430 Motor Circuits, and Controllers. is the main *NEC* motor Article. *Page 324, Table 430.5 Other Articles.* gives *NEC* Articles and Sections that discuss the application of motors and motor controllers in specific equipment.

157) Often, *NEC* table values of a motor's full-load current are used rather than the motor's nameplate Full-Load Amperage (FLA) rating values for determining the ampacity of conductors and the ampere ratings of switches and branch-circuit Short-Circuit and Ground-Fault Protection. Where is this stated in the *NEC*? (Information is required from several Sections and Tables.)

Page 324, Section 430.6 Ampacity and Motor Rating Determination., (A) General Motor Applications., (1) Table Values. --- Yes, with some exceptions, the motor current values given in the following *NEC* tables shall be used rather than the motor's nameplate Full-Load Amperage (FLA). See *Page 349, Table 430.247 Full-Load Current in Amperes, Direct-Current Motors.; Page 350, Table 430.248 Full-Load Currents in Amperes, Single-Phase Alternating-Current Motors.; Page 350, Table 430.249 Full-Load Current, Two-Phase Alternating-Current Motors (4-Wire).; or Page 351, Table 430.250 Full-Load Current, Three-Phase Alternating-Current Motors.*

158) Where does the *NEC* describe motor nameplate data?

Page 325, Section 430.7 Marking on Motors and Multimotor Equipment.

159) What nameplate data should be on a motor controller?

Page 326, Section 430.8 Marking on Controllers. --- Manufacturer's name, voltage, current or horsepower, and Short-Circuit rating.

160) Where does the *NEC* require motors to be protected against liquids and dust? (Information is required from several Sections.)

Page 327, Section 430.11 Protection Against Liquids. and *Page 328, Section 430.16 Exposure to Dust Accumulations.*

161) What percent of motor full-load nameplate current should a separate Overload Device be rated for the following?
a)   2 horsepower, Service Factor 1.0, temperature rise 40°C
b)   2 horsepower, Service Factor 1.15, temperature rise 30°C
c)   2 horsepower 2 horsepower, Service Factor 1.15, temperature rise 40°C
d)   1/2 horsepower, Service Factor 1.15, temperature rise 40°C
(Information is required from several Sections.)

Page 332, Section 430.32 Continuous-Duty Motors., (A) More Than 1 Horsepower., (1) Separate Overload Device. --- a) Should be 115%. b) Should be 125%. c) Should be 125%.

Page 332, Section 430.32 Continuous-Duty Motors., (B) One Horsepower or Less, Automatically Started., (1) Separate Overload Device. --- d) Should be 125%.

162) What size dual element fuse is required by a 3-phase induction motor?

Page 335, Table 430.52 Maximum Rating or Setting of Motor Branch-Circuit Short-Circuit and Ground-Fault Protective Devices. --- The dual element fuse should have a rating of 175% of its full load current.

Page 335, Section 430.52 Rating or Setting for Individual Motor Circuit., (C) Rating or Setting., (1) In Accordance with Table 430.52., Exception No. 1: --- If the calculated current does not fall on a standard fuse rating, the next larger fuse may be chosen.

163) May three ¼ horsepower 120 volts single-phase induction Thermal Overload Protected motors be powered through the circuit breaker of one branch-circuit?

Page 337, Section 430.53 Several Motors or Loads on One Branch-Circuit., (A) Not Over 1 Horsepower. --- Yes, if the ratings of the circuit breaker are not exceeded.

164) Is a motor controller required to have its own Short-Circuit and Ground-Fault Protective Devices?

Page 338, Section 430.72 Overcurrent Protection., (A) General. --- If it is tapped off a branch-circuit a Short-Circuit and Ground-Fault Protective Device are not necessary.

165) What protection is allowable on a 120/24 V 40 volt-amperes transformer that is part of a motor controller and in its motor controller enclosure?

Page 339, Section 430.72 Overcurrent Protection., (C) Control Circuit Transformer., (3) Less Than 50 Volt-Amperes. --- It may be protected by a primary Overcurrent Device, Impedance Limiting, or an inherent protective means.

166) What size motor controller is needed for a 50 horsepower 460 volts 3-phase induction motor?

Page 340, Section 430.83 Ratings., (A) General., (1) Horsepower Ratings. --- The motor controller has to be rated for at least 50 horsepower at 460 volts.

167) May motor controllers be used on voltages other than their rated voltage?

Page 341, Section 430.83 Ratings., (E) Voltage Rating. --- Yes, they may be used at voltages lower than their rated voltage.

168) Is an individual disconnecting means required for each motor controller?

Page 343, Section 430.102 Location., (A) Controller. --- Yes, an individual disconnecting means shall be provided for each controller. It shall be located in sight of the controller.

169) May a motor controller disconnect also serve as a motor disconnect?

Page 343, Section 430.102 Location., (B) Motor., (2) Controller Disconnect. --- Yes, if it is within sight of the motor location and driven machinery.

170) Where does the *NEC* state that motor and motor controller disconnecting means shall plainly indicate open (off) and closed (on) positions?

Page 343, Section 430.104 To Be Indicating.

171) If the rated input current of an inverter is 100 amperes, what is the minimum ampacity of the conductors supplying it?

Page 346, Section 430.122 Conductors — Minimum Size and Ampacity., (A) Branch/Feeder Circuit Conductors. --- The minimum ampacity is 1.25 x 100 = 125 amperes.

172) May overload equipment be included within power conversion equipment?

Page 346, Section 430.124 Overload Protection., (A) Included in Power Conversion Equipment. --- Yes.

173) Is motor over-temperature protection required if the motor is powered by an adjustable speed drive?

Page 346, Section 430.126 Motor Overtemperature Protection., (A) General. --- Yes, unless the motor is rated to run over the full application speed range at nameplate rated current.

174) A full-wave rectifier supplies a 120 volts dc 1 horsepower motor in a 30°C or less environment.
a)   What is the motor's full-load current?
b)   What size THHN copper conductors could be used here?
          (Information is required from a Sections and Tables.)

a) _Page 349, Table 430.247 Full-Load Current in Amperes, Direct-Current Motors._ --- From the table the motor's full-load current is 9.5 amperes.

b) _Page 329, Section 430.22 Single Motor., (A) Direct-Current Motor-Rectifier Supplied., (2)._ --- The wiring from the rectifier to the motor shall be capable of 150 percent of the motor's full-load current. The wiring should be capable of 1.5 x 9.5 = 14.2 amperes.

Page 161, Table 310.15(B)(16) (formerly Table 310.16) Allowable Ampacities of Insulated Conductors Rated Up to and Including 2000 Volts, 60°C Through 90°C (140°F Through 194°F), Not More Than Three Current-Carrying Conductors in Raceway, Cable, or Earth (Directly Buried), Based on Ambient Temperature of 30°C (86°F)*. --- This table indicates that 16 AWG copper conductor could be used.

Page 329, Section 430.22 Single Motor., (G) Conductors for Small Motors. --- This indicates that the conductor size should not be less than 14 AWG copper conductor. 14 AWG copper conductor is the final choice.

175) What gauge SOOW copper 16/3 flexible cord should be used to power a single-phase 1 horsepower 115 V table saw motor? The motor's nameplate full-load amperage is 11 amperes. The table saw will need to be able to operate continuously in a 40 °C environment. The circuit will be protected by a dual element fuse. (Information is required from several Sections and several Tables.)

Page 350, Table 430.248 Full-Load Currents in Amperes, Single-Phase Alternating-Current Motors. --- The table gives the motor's full-load current as 16 amperes. Notice this is greater than the motor's nameplate full-load amperage.

Page 335, Table 430.52 Maximum Rating or Setting of Motor Branch-Circuit Short-Circuit and Ground-Fault Protective Devices. --- A dual element fuse should have a rating of 1.75 times the full load current. 1.75 x 16 = 28.0 amperes.

Page 335, Section 430.52 Rating or Setting for Individual Motor Circuit., (C) Rating or Setting., (1) In Accordance with Table 430.52., Exception No. 1:--- If the calculated current does not fall on a standard fuse rating, the next larger fuse may be chosen.

Page 96, Section 240.6 Standard Ampere Ratings., (A) Fuses and Fixed-Trip Circuit Breakers. --- The next largest standard size is 30 amperes.

SOOW flexible cord is on Page 269, Table 400.5(A)(1) Allowable Ampacity for Flexible Cords and Cables [Based on Ambient Temperature of 30°C (86°F). See 400.13 and Table 400.4. The table's capacity in column B[b] for the SOOW 8 AWG cable is 40 amperes. at 30 °C.

The ampacity at 30 °C needs to have an "ampacity adjustment" to find its ampacity at 40 °C. SOOW cord, according to vendor literature, is rated to 90 °C. Using the 90 °C column on Page 158, Table 310.15(B)(2)(a) Ambient Temperature Correction Factors Based on 30°C (86°F). From this the correction factor is .91. The adjusted ampacity is .91 x 40 = 36.4 amperes.

Since the 36.4 amperes is greater than the fuse's 30 amperes rating, 8AWG SOOW cord may be used here.

176) What are the full-load and locked rotor currents for a 7½ horsepower, 230 volts single-phase, code letter G induction motor? (Information is required from several Tables.)

Page 350, Table 430.248 Full-Load Currents in Amperes, Single-Phase Alternating-Current Motors. --- This table indicates the full-load current of the motor is 40 amperes.

Page 326, Table 430.7(B) Locked-Rotor Indicating Code Letters. --- This table indicates the kilovolt-amperes per horsepower are 5.6–6.29. From this the kilovolt-amperes will range from 5.6 x 7.5 = 42 kilovolt-amperes to 6.29 x 7.5 = 47.2 kilovolt-amperes. The locked rotor current is 42,000/230 = 182.6 A to 47,200/230 = 205.2 amperes.

177) What current rating should a thermal protector have on a 120 volts 1½ horsepower single-phase motor? (Information is required from a Section and Table.)

Page 350, Table 430.248 Full-Load Currents in Amperes, Single-Phase Alternating-Current Motors. --- Full-load current is 20 amperes.

Page 332, Section 430.32 Continuous-Duty Motors., (A) More Than 1 Horsepower., (2) Thermal Protector., --- Thermal protector rating should be not more than 1.56 x 20 = 31.2 amperes.

178) What gauge copper THHN wire is needed to continuously supply three 230 volts 3-phase induction motors? The motors are 10 horsepower, 3 horsepower, and 1 horsepower. Assume an ambient temperature of 30°C. (Information is required from a Section and several Tables.)

Page 351, Table 430.250 Full-Load Current, Three-Phase Alternating-Current Motors. --- From the table the full-load currents are:

    10 horsepower   28.0   amperes
     3 horsepower   9.6 amperes
     1 horsepower   4.2 amperes

Page 330, Section 430.24 Several Motors or a Motor(s) and Other Load(s). --- Following the calculation method of this Section. The conductor should have an ampacity not less than 1.25 x (28) + (9.6 + 4.2) + 1 x (0) + 1.25 x (0) = 48.8 amperes

Page 161, Table 310.15(B)(16) (formerly Table 310.16) Allowable Ampacities of Insulated Conductors Rated Up to and Including 2000 Volts, 60°C Through 90°C (140°F Through 194°F), Not More Than Three Current-Carrying Conductors in Raceway, Cable, or Earth (Directly Buried), Based on Ambient Temperature of 30°C (86°F).* --- This table indicates that 8 AWG could be used.

179) What minimum ampere rating is required for an inverse-time circuit breaker used on a 50 horsepower 460 volts 3-phase induction motor? Note: The minimum ampere rating is not the same as overcurrent protection rating. Here ampere rating indicates the capacity of the circuit breaker to carry continuous current. (Information is required from a Section and Table.)

Page 351, Table 430.250 Full-Load Current, Three-Phase Alternating-Current Motors. --- The table indicates this motor has a full-load current of 65 amperes.

Page 344, Section 430.110 Ampere Rating and Interrupting Capacity., (A) General. --- Using this Section the circuit breaker shall be rated for at least 1.15 x 65 = 75 amperes at 460 volts. Alternatively, the circuit breaker could be rated for at least 50 horsepower at 460 volts.

180) The *NEC* states that it covers the installation of all transformers, but then gives exceptions. What are some of the exceptions?

Page 360, *Section 450.1 Scope.*, *Exception No. 1 to 8.* --- Some examples are: current transformers, dry-type transformers that are part of a piece of apparatus, such as a computer power supply, sign lighting transformers, and small low power & voltage transformers.

181) What conductor sizes and Overcurrent Protection are needed by an air conditioner that has the following electrical nameplate data? Assume the temperature is no more than 30°C, the conductors are THHN, and there are no more than two current carrying conductors in a raceway. (Information is required from a Section and Table.)

Nameplate of an Air Conditioner that uses an Hermetic Refrigerant Motor-Compressor				
Volts AC	**Min Volts**	**Max Volts**	**Phase**	**Frequency**
208/230	197	253	1	60 Hertz

Minimum Circuit Ampacity	**Max Fuse or Circuit Breaker Amperes**			
35.7	50			

Compressor				
RLA Amperes	**LRA Amperes**	**Horsepower**		
27.1	144	-		

Fan				
FLA Amperes	**LRA Amperes**	**Horsepower**		
1.9	3.7	1/3		

<u>Page 354, Section 440.6 Ampacity and Rating., (A) Hermetic Refrigerant Motor-Compressor.</u> --- The *NEC* states that nameplate data should be used to determine conductor ampacity and Overcurrent Protection rating. From this the conductor ampacity should be at least the 35.7 amperes of the nameplate.

<u>Page 161, Table 310.15(B)(16) (formerly Table 310.16) Allowable Ampacities of Insulated Conductors Rated Up to and Including 2000 Volts, 60°C Through 90°C (140°F Through 194°F), Not More Than Three Current-Carrying Conductors in Raceway, Cable, or Earth (Directly Buried), Based on Ambient Temperature of 30°C (86°F)*.</u> --- The table shows that 10 AWG THHN conductors could be used here.

A 40 amperes fuse or circuit breaker could protect this. However, given that the nameplate states that an up to 50 amperes Overcurrent Device could be used, another possibility would be 8 AWG conductors with a 50 amperes Overcurrent Device.

182) What size fuses should a single-phase 100 kilovolt-amperes 2400/240 volts distribution transformer use? Assume an unsupervised location and a percent transformer impedance of less than 6%. (Information is required from a Section and Table.)

Page 361, Table 450.3(A) Maximum Rating or Setting of Overcurrent Protection for Transformers Over 1000 Volts (as a Percentage of Transformer-Rated Current). --- Rated primary current is 100,000/2400 = 41.7 amperes. Rated secondary current is 100,000/240 = 417 amperes. From the table the primary fuse should be 3 x 41.7 = 125 amperes and the secondary fuse should be 1.25 x 417 = 521 amperes. Note 1 of Table 450.3(A) states that if the calculated fuse value is not a standard value, the next larger fuse may be used. Page 97, Section 240.6 Standard Ampere Ratings., (A) Fuses and Fixed-Trip Circuit Breakers. gives the appropriate standard fuse sizes as 125 amperes and 600 amperes. Note: 600 amperes is the next larger standard size.

183) What size fuses should a single-phase 10 kilovolt-amperes 480/120 volts transformer use? Assume the transformer circuit does not meet the "only primary needs to be protected" specifications of Page 99, Section 240.21 Location in Circuit., (C) Transformer Secondary Conductors. (Information is required from several Tables and a Section.)

Page 362, Table 450.3(B) Maximum Rating or Setting of Overcurrent Protection for Transformers 1000 Volts and Less (as a Percentage of Transformer-Rated Current). --- Rated primary current is 10,000/480 = 20.8 amperes. Rated secondary current is 10,000/120 = 83 amperes. From the table the primary fuse should be 2.5 x 20.8 = 52 amperes and the secondary fuse should be 1.25 x 83 = 104 amperes. According to Table 450.3(B), rounding up is allowed on the secondary fuse, but not on the primary fuse. Standard fuses lower than 52 and higher than 104 amperes should be selected. Page 97, Section 240.6 Standard Ampere Ratings., (A) Fuses and Fixed-Trip Circuit Breakers. indicates that 50 and 110 amperes standard fuses could be used.

184) Do Overcurrent Devices for transformers protect their conductors?

Page 361, Section 450.3 Overcurrent Protection. and Page 94, Section 240.4 Protection of Conductors. --- They may. With care and attention to both NEC Sections it is possible that the same Overcurrent Device would protect both transformer and its conductors.

185) What transformers are required to have a disconnecting means? In what locations are the disconnects required?

Page 365, Section 450.14 Disconnecting Means. --- Other than Class 2 or 3 transformers, disconnecting means are required by all transformers. The location of the disconnect may be within sight of the transformer or in a remote location. If a remote location is used, it must be marked on the transformer and the disconnect must be lockable.

186) What is a Class I Division 1 location?

Page 384, Section 500.5 Classifications of Locations., (B) Class I Locations., 1) Class I, Division 1. --- A location where ignitable concentrations of gases or liquid-produced vapors may exist under normal conditions.

187) What does Class III refer to?

Page 385, Section 500.5 Classifications of Locations., (D) Class III Locations. --- This is a hazardous location classification that has easily ignitable fibers or combustible flyings. It should not be confused with Class 3, a circuit classification.

188) In a hospital anesthetizing area, what is the highest voltage allowed between conductors before connection to an equipment grounding conductor is required?

Page 477, Section 517.62 Grounding., Exception. --- Not more than 10 volts.

189) For the purposes of the bonding of equipotential surfaces in agricultural buildings, are chicken barns and cattle barns required to meet the same bonding requirements?

Page 504, Section 547.10 Equipotential Planes and Bonding of Equipotential Planes. --- For the purposes of equipotential bonding, chickens are not considered to be livestock and do not need barns that meet the same bonding requirements.

190) Does the *NEC* allow boats to be supplied power via temporary wiring?

Page 538, Section 555.13 Wiring Methods and Installation., (A) Wiring Methods., (3) Temporary Wiring. --- Generally, no.

191) What is the longest time that temporary holiday wiring is permitted?

Page 540, Section 590.3 Time Constraints., (B) 90 Days. --- 90 days.

192) Are GFCIs required for temporary construction wiring?

Page 541, Section 590.6 Ground-Fault Protection for Personnel. --- Yes.

193) The *NEC* has an Article on Information Technology Equipment. What is the NFPA document the *NEC* refers to for the protection of information technology equipment?

Page 584, Section 645.1 Scope., Informational Note: --- *NFPA 75-2013, Standard for the Protection of Information Technology Equipment.*

194) When must abandoned cables be removed from information technology equipment systems?

Page 586, Section 645.5 Supply Circuits and Interconnecting Cables., (G) Abandoned Supply Circuits and Interconnecting Cables. --- Accessible portions of the cables must be removed unless they are contained in raceways.

195) Is bonding required for swimming pool underwater lighting?

Page 613, Section 680.26 Equipotential Bonding., (B) Bonded Parts., (4) Underwater Lighting. --- Generally yes, although listed low-voltage lighting systems with non-metallic forming shells do not require it.

196) A transfer switch should be used when connecting an optional standby generator to a house's electrical system. Where does the *NEC* state this?

Page 662, Section 702.5 Transfer Equipment.

197) What are Class 1 circuits? (Information is required from several Sections.)

Page 673, Section 725.2 Definitions., Class 1 Circuit. and _Page 674, Section 725.41 Class 1 Circuit Classifications and Power Source Requirements._ --- Class 1 circuits are remote-control, signaling, and power-limited circuits. They are between the load side of an Overcurrent Protective Device and load equipment. There are two types of Class 1 circuits. The first is called "Power-Limited". It is supplied from a source that has a rated output of not more than 30 volts and 1000 volt-amperes. The second is called "Remote-Control and Signaling". It is limited to 600 volts but there is no power limit.

198) What are Class 2 and Class 3 circuits?

Page 673, Section 725.2 Definitions., Class 2 Circuit., Class 3 Circuit. --- Class 2 circuits control things like doorbells and thermostats. They should not be able to cause fire or electric shock. Class 3 circuits are similar to Class 2 except that they may be able to start a fire. Class 3 should not be confused with Class III, a hazardous location classification.

199) Power or lighting conductors may not be in the same raceway as Class 2 or Class 3 conductors. Where is this stated in the _NEC_?

Page 680, Section 725.136 Separation from Electric Light, Power, Class 1, Non-Power-Limited Fire Alarm Circuit Conductors, and Medium-Power Network-Powered Broadband Communications Cables.

200) Must fire alarm circuits be separated from power and lighting circuits?

Page 694, Section 760.136 Separation from Electric Light, Power, Class 1, NPLFA, and Medium-Power Network-Powered Broadband Communications Circuit Conductors. --- Power Limited Fire Alarm circuits must be separated from circuits carrying power.

201) Determine the minimum size for EMT conduit that contains eight stranded 12 AWG THHN cables. (Information is required from several Tables.)

Page 756, Table 1 Percent of Cross Section of Conduit and Tubing for Conductors and Cables. --- The table indicates that since there are more than 2 conductors, the total cross-sectional area of the conductors should not be more than 40% of the area of the conduit.

Page 763, Table 5 Dimensions of Insulated Conductors and Fixture Wires., --- The approximate area per conductor is 8.581 mm² (.0133 in.²). With 8 conductors the total area is 8 x 8.581 = 69 mm² (.106 in.²).

Page 757, Table 4 Dimensions and Percent Area of Conduit and Tubing (Areas of Conduit or Tubing for the Combinations of Wires Permitted in Table 1, Chapter 9) --- Metric Designator 16 (Trade Size ½) EMT has a 40% useable cross-sectional area of 78 mm² (.122 in.²) 40% useable means 60% of the conduit cross-section will be empty.

The answer is Metric Designator 16 (Trade Size ½) EMT.

202) What is compact stranding?

Page 794, Table C.1(A) Maximum Number of Conductors or Fixture Wires in Electrical Metallic Tubing, Definition: --- Compact stranding applies to multi-strand conductors. The individual strands on a compact stranded conductor are crushed together so that there are no spaces between the strands.

4.0   QUESTIONS THE *NEC* DOES NOT ANSWER

There are questions and topics that it would seem the *NEC* would cover, but does not. Some of these follow.

1) Does the *NEC* require licensed electricians or electrical engineers for electrical work?

The *NEC* makes no mention of licensing, electricians, or electrical engineers. However, they are mentioned on <u>Pages 856 to 863, Informative Annex H Administration and Enforcement.</u> This section is printed with the *NEC*, but is not considered to be part of the official *NEC*. In this section, the *NEC* board requires electrician membership. Also it mentions that an electrical engineering degree is one of several possible ways of qualifying to be an Electrical Inspector.

The *NEC* does make a distinction between "qualified persons" and "unqualified persons". It defines a "qualified person" on <u>Page 33, Article 100 Definitions., I General.</u> The *NEC* often describes locations where "qualified persons" may work with circuits and equipment that "unqualified persons" are not allowed to access.

2) Do the terms "bus duct" and "Romex" appear in the *NEC*? If not, what terms are used in their place?

A keyword search of the *NEC* PDF version shows that "bus duct" and "Romex" are not used. In their place the terms busway and type NM cable (<u>N</u>on-<u>m</u>etallic cable) are used.

3) Does the *NEC* allow touching of conductors to test if they are energized?

The *NEC* does not discuss testing methods or electrical safety. The <u>NFPA 70E Standard for Electrical Safety in the Workplace.</u> states that contact should be avoided for ac voltages greater than 50 volts and prohibited for voltages 301 volts or greater.

4) Does the *NEC* specify the percent voltage drop from the service to the load for voltages more than 600 volts?

No. It makes recommendations for circuits of 600 volts or below in <u>Page 57</u>, <u>Section 210.19 Conductors — Minimum Ampacity and Size.</u>, <u>(A) Branch-Circuits Not More Than 600 Volts.</u>, <u>Informational Note No. 4:.</u>, but does not make them for above 600 volts.

5) What do the terms "ampacity adjustment" and "ampacity correction" mean in the *NEC*?

They refer to a re-rating of a conductor's ampacity due to ambient temperature or closeness to other conductors. The terms are not defined in the *NEC*.

6) What is the definition of "double insulation"?

Although the *NEC* uses the words "double insulation" and "double insulated", it does not define them. A double insulated electrical appliance is designed so that it can be operated safely without a connection to earth ground. This is usually done by using two (double) layers of insulation around live parts.

7) What is the definition of "Short-Circuit"?

The *NEC* does not define "Short-Circuit" even though it uses the term many times. UL Standard 489 defines it as "An abnormal connection (including an arc) of relatively low impedance, whether made accidentally or intentionally, between two points of different potentials."

8) What does the *NEC* specify for the installation of lightning protection systems?

<u>Page 126</u>, <u>Section 250.106 Lightning Protection Systems.</u> --- The NEC does mention lightning protection; however, for details refers the reader to the NFPA 780, Standard for the Installation of Lightning Protection Systems.

9) How many NM conductors may be properly put through a bore hole in a house's wall stud?

Page 143, _Section 300.4 Protection Against Physical Damage._ --- The _NEC_ only states that the conductors must be protected against physical damage. There is nothing in the _NEC_ that states how many NM conductors may be run through a bore hole. However, your local municipality may have their own standard on this. A reasonable estimate could be made by assuming the same number of conductors as could be put into a raceway of the same hole diameter. See _Page 756_, _Chapter 9 Tables._, _Table 1 Percent of Cross Section of Conduit and Tubing for Conductors and Cables._

10) What is the definition of "Service Factor"?

The _NEC_ does not define "Service Factor", but it does use the term. Service Factor is a measure of the overload capacity of a motor. A 1.15 SF (_S_ervice _F_actor) means the motor can deliver up to 15% horsepower beyond its rating without being damaged.

5.0 OTHER SOURCES OF INFORMATION

5.1 REFERENCES

Holt, Mike, 2014: *Understanding the NEC*, Volume 1, Articles 90 to 480, price $65.00 and *Understanding the NEC*, Volume 2, Articles 500 to 820, Mike Holt Enterprises, Inc., price $52.50. These books illustrate the *NEC* with excellent hands-on examples and graphical explanations.

Miller, Charles R., 2011. *Illustrated Guide to the National Electrical Code,* 5th Ed., Clifton Park, NY: Delmar Cengage Learning, price $101.95. It contains many good explanatory drawings. Mastering this book would assure passage of many electrician licensing exams.

NFPA 70: National Electrical Code,2014 Ed., Quincy, MA: National Fire Prevention Association, price $89.50.

NFPA 70: 2014 Handbook, Quincy, MA: National Fire Prevention Association, price $165.50. This handbook includes the entire *NEC* and also provides examples, diagrams, photos, and explanations. Some prefer to use it rather than the *NEC*.

5.2 USEFUL WEBSITE

Mike Holt Enterprises, Inc. The website has a great deal of information on the *NEC* and related subjects. Much of the information is free.
http://www.mikeholt.com/

5.3 NFPA

National Fire Prevention Association
One Batterymarch Park
Quincy, Massachusetts 02169-7471
1 617 770-3000
http://www.nfpa.org/

5.4 WHERE TO PURCHASE THE *NEC*

The NFPA will sell you the *NEC* directly. NFPA members receive a discount. The *NEC* is also available at online bookstores.

5.5 *NEC* CLASSES

For-a-fee *NEC* classes are available from the NFPA. Currently the NFPA offers four-day classes on the *NEC* at different locations in the U.S. The cost is $1,345. The NFPA will teach *NEC* classes at your employer's site. For an extra fee they offer a *NEC* training certification examination following the training.

The NFPA offers on-line self-guided *NEC* courses. They have it in course modules. Costs are from $60 to $324 per course.

For-a-fee *NEC* classes are taught by other companies. Do a search on the internet for information.

9.0 *NEC* ABBREVIATIONS USED IN THIS BOOK

AC = <u>A</u>rmored <u>C</u>able

AFCI = <u>A</u>rc-<u>F</u>ault <u>C</u>ircuit <u>I</u>nterrupter

AHJ = <u>A</u>uthority <u>H</u>aving <u>J</u>urisdiction

AWG = <u>A</u>merican <u>W</u>ire <u>G</u>auge

EMT = <u>E</u>lectrical <u>M</u>etallic <u>T</u>ubing

FLA = <u>F</u>ull-<u>L</u>oad <u>A</u>mperage

GFCI = <u>G</u>round-<u>F</u>ault <u>C</u>ircuit <u>I</u>nterrupter

IC Luminaire = Special light bulb housing designed for direct contact with thermal insulation

LRA = <u>L</u>ocked <u>R</u>otor <u>A</u>mperage, applies to motors when rotor is stopped and full voltage applied

MC = <u>M</u>etal-<u>C</u>lad Cable

NM = <u>N</u>on-<u>M</u>etallic Sheathed Cable

NPLFA = <u>N</u>on-<u>P</u>ower-<u>L</u>imited <u>F</u>ire <u>A</u>larm

PVC = Rigid <u>P</u>oly<u>v</u>inyl <u>C</u>hloride Conduit

RLA = <u>R</u>ated <u>L</u>oad <u>A</u>mperage

RTRC-XW = <u>R</u>einforced <u>T</u>hermosetting <u>R</u>esin <u>C</u>onduit in a location subject to physical damage

SE = <u>S</u>ervice <u>E</u>quipment

SOOW = Service cord with an Oil resistant jacket, Oil resistant insulation, and Weather resistance

THHN = Thermoplastic High Heat Resistant Nylon Coated Wire

UF = Underground Feeder and Branch-Circuit Cable

USE = Underground Service Equipment

www.ingramcontent.com/pod-product-compliance
Lightning Source LLC
Chambersburg PA
CBHW080555220326

41599CB00032B/6492